Only in America

From the South Bronx to Silicon Valley

My Journey Through Life
and Service to this Nation and Beyond

A memoir by

Bernard P. Marcus

Only in America
From the South Bronx to Silicon Valley

© 2017 Bernard P. Marcus

ISBN: 978-1-61170-258-3

Published by:

 Robertson Publishing™
www.RobertsonPublishing.com

Printed in the USA and UK on acid-free paper.
To purchase additional prints of this book go to:

amazon.com
barnesandnoble.com

Table of Contents

Prologue

Whenever anything good happened to her children, my mother would exclaim with pride and joy, "Only in America!"

I am an American-born son of Jewish immigrants from Eastern Europe. My maternal grandparents and my parents emigrated in the early 1900s and settled in the slums of Manhattan and the Bronx. I was born in 1922 and spent my growing up years in the South Bronx.

I survived and escaped the Bronx after World War II broke out. I served as an officer in the Counter Intelligence Corps (CIC) in Germany. The most gratifying part of that service was helping Jews leave Germany on their way to Palestine. The CIC was my introduction to the CIA, which I was invited to join but declined. However, the CIA was instrumental in providing me with opportunities for an exciting career, for which I am eternally grateful.

The highlights of *Only in America* include the development and testing of an aerial photo–reconnaissance capability for the Army during the Korean War, followed by the development of the first drone reconnaissance system. Then came the camera systems for the U-2 airplane, culminating in the camera system for the first U.S. satellite. All of these were sponsored by or run by the CIA, which arranged for me to be responsible for the photographic systems. *Only in America* could a first-generation Jewish boy have these opportunities.

I married Molly Cohen on October 27, 1946. We were married until her death on January 26, 1992. Ruthie Howard and I were married on March 28, 1993.

This memoir is a compilation of snippets, which span my life from the slums of the Bronx to our "cruise ship" — the Vi, a retirement community in Palo Alto. Many of these snippets were written in response to a more recent event; those events are referenced at the beginning of the snippet.

Growing Up in the Bronx

The Early Years

I was born in the Bronx in New York City on May 14, 1922. My mother's maiden name was Betty Lerner. She came to this country from Galicia, a Polish province of the Austro-Hungarian Empire in 1908 when she was eight years old. She said she was from Austria. My father's name was David Marcus. He migrated in 1912 at age sixteen from Romania. He fought in World War I and was wounded by shrapnel and mustard gas. He received the Purple Heart medal, of which he was very proud. He and my mother were married July 3, 1921. They would never talk about how they met. We lived in a ground-floor apartment at 643 East 138th St. until I was seven years old. Some of the memories that have remained with me of that time are:

- In 1925, my sister Millie was born in our apartment—screaming.

- My father's sister and her family, the Altarescus, lived down the block and had a retail radio store in the same block. We spent a great deal of time with them, especially on Sundays and holidays. We had lunches and dinners (suppers, as we called them) together.

- St. Ann's Catholic Church, was down the block from where we lived. At Easter time my parents kept me indoors as the kids came running out of services. They wanted to beat up Jewish kids because they were taught we killed Christ.

To encourage people to move into an apartment during the Great Depression, landlords would give tenants the first and last months' rent-free in exchange for signing a one-year lease. This was only effective on September 1 so that children could get signed up for a new school, if necessary. My parents took advantage of that big money saver. We moved from 138th St., first to 235 Cypress Avenue September 1, 1929, and then to a new apartment building in the Bronx at 440 Jackson Avenue September 1, 1930.

My father loved this second-floor apartment because our living room and my parent's bedroom faced St. Mary's Park. It was probably the best view from an apartment house in the South Bronx. Also the Jackson Avenue apartment was relatively new. The rent was $30 per month, with the first month free after my father signed a year's lease.

I had two close friends, one Jewish (Sam) and one Italian (Louis Claps). Of the two, I kept in touch with Louis until I was about 40 years old. Of all the kids in the neighborhood, Louis and I were the only two to go to college. Louis went to Manhattan College and then into the priesthood.

My recollections of growing up on Jackson Avenue center around walking to school, playing stickball, handball on the side of the apartment buildings, box ball on the sidewalk, marbles (immies, as we called them) and having rock fights with kids from other neighborhoods. We had wet and bitter cold winters and hot steamy summers.

In the hot summers, we would sleep outside on the fire escape. Baseball was big. We knew the hitting percentages of the all players. I was a Yankee fan because we lived in the Bronx. I would go to games with my older cousins until I was 12. Then I was allowed to go to double headers on holidays by myself. I took a trolley car on 149th St. to the Grand Concourse. The fare was five cents. Then I walked six blocks to Yankee Stadium. Bleacher seats were 25 cents, and I would sit in right field behind Babe Ruth.

Fridays were intense. My mother started cooking early Friday morning for the weekend. Since she was orthodox, no cooking was permitted from Friday sundown until Saturday sundown. On those Fridays that she didn't cook, we travelled by subway to my grandparents' apartment on the Lower East Side of New York City and stayed until Saturday night.

My younger sister Elaine was born in 1936, when I was 14 years old. It was pretty crowded in our three-room apartment. I slept in the living room. I was going to James Monroe High School by then, and the only way that I could have it quiet enough to do homework was to rock Elaine to sleep while my mother served dinner to my father. In 1936 we were in the midst of the Great Depression. Luckily, my father, an optician, always had a job. His wages were pretty low, about $20 per week, but we seemed to do OK. He worked six days a week for as long as I can remember, except for one period of time when he only worked on weekends. We were never hungry, and in the summer, we rented a bunga-low in Rockaway Beach with money my mother was able to stash away.

I graduated from high school in January 1938, and my father sent me to work on a dairy farm in Accord, New York, because he said I was too young to start college. I was happy to leave the farm in August and started at City College of New York in Sep-tember 1938.

At my graduation ceremony at City College in June 1942, I received a medal from the United Daughters of the Confederacy for having written the best story of the South by an ROTC Gradu-ating Cadet. I later found out that this was very disturbing to the Daughters of the Confederacy, because City College was known as *The Communist College* in the country.

Early Memories Of My Father

I never called my father *Dad*; I always called him *Pop*.

The earliest thing I remember happened in January 1925. I was almost three years old. We were living on 138th Street. I remember my father carrying me outside in his arms, telling me that it was midday and the sun would disappear slowly and it would get dark. He let me look at the moon covering the sun through a small hole in a newspaper and pointed out the birds flying around very confused. It was the first and last total eclipse of the sun I have witnessed.

Every day, my father would take a long walk to the 3rd Avenue elevated train (called the 3rd Avenue El) to go to work.

Supper was always at 6 p.m. My mother, my sister Millie and I had to finish supper before my father came home from work, just before 7. He liked to eat quietly by himself while listening to Gabriel Heatter and then *Amos and Andy* on the radio. During the Depression, Gabriel Heatter was great. He always started his news program with, "Good evening, everyone; there is good news tonight." I was required to listen to Gabriel Heatter so I became familiar with the news. My mother and sister were often in the living room reading.

When we lived on Jackson Ave., every work night after supper, but never on Sunday, Pop took me for a two-mile walk around the Park. As soon as we left the apartment house, he lit up a *Between the Acts* cigar. He would often tell me, "I never smoke in the house. The smoke is no good for your mother or you kids. Remember that when you grow up." Then he would tell me what went on at work and the news of the day, which he had read in the *Journal American* on the El. This went on until Elaine was born.

On Sundays he would take an afternoon nap. My mother insisted that there be absolute silence at that time, since he worked ten hours a day, six days a week. That was her way of taking care of him.

He also took good care of my mom. Every morning, at about 6:30, he would walk to the bakery and bring home kaiser rolls and bread, either rye or pumpernickel. He then prepared a pot of hot cereal for breakfast, which he made sure my sister Millie and I ate. He then made himself a cheese sandwich for his lunch. He would leave for work about 8, after he awakened my mom and kissed her goodbye.

Then there were the parades. On Decoration Day (now Memorial Day) and Armistice Day (now Veterans Day), my father would put on his army cap, pin on his World War I medals and join his buddies of the Order of the Purple Heart for the big parade down Fifth Ave. My mother, my sisters and I would always proudly watch. In 1938, I went downtown with him to parade behind the Purple Heart group as a member of the City College ROTC. My mother and my sisters Millie and Elaine would watch and cheer from the steps of St. Patrick's Cathedral. We were in the same parades through Decoration Day 1942, after which I left for the Army and World War II.

Today is Valentine's Day 2012. Every Valentine's Day, my father would bring my mother a Whitman's Chocolate Sampler box. She loved it because on the inside of the cover of the box the names and locations of the chocolates were listed.

Fond Memories.

Memories Of Kindergarten through Sixth Grade

These are some of the memories of events that took place around P.S. 65, the grammar school I attended. The school was on Cypress Avenue and 141st Street in the Bronx. From our apartment on 138th Avenue, my mother took me to school on my first day of kindergarten. After that day I walked by myself or with some buddies because this was considered a safe walk. From our apartment building on Jackson Avenue and 145th Street, it was a six-block walk.

Apartment Building at 440 Jackson Avenue

As best I can remember, the following took place in the grades I've indicated:

- First grade—The teacher handed out drawing paper with a scene to be colored. She looked at what I had done and said, "This is terrible, take this home to your mother with this note." The note read, "Teach your son the colors." My mother told me, "Don't worry, I'll help you, and you'll learn them." Of course, I had no idea then that I was colorblind.

- First through sixth grade—Every Monday morning, after we recited the Pledge of Allegiance, we had to hold out our hands as the teacher inspected them. She was looking for dirty, uncut nails. If you were one of those she identified, it was off to the nurse's office where you had to scrub your hands and nails and have your nails cut. The nurse would then give you a note to take home to be returned with your mother's signature.

- First through sixth grade—At ten o'clock we had a break, and each of us was given a small container of milk that we had to drink.

- Third grade—One day our teacher told us that the principal was concerned that many of us were too skinny and were not having a good breakfast before we came to school. She then pointed to a plump girl in our class whom she said looked really healthy. She asked her to stand up and tell the class what she had for breakfast. She replied, "A roll and budder, and a glass of cawfee."

- Second through sixth grade—Penmanship. We were given handwriting worksheets and were required to write in cursive. I can still read what I write.

A Typical Week at Home, Ages Eight through Sixteen

As much as my mother loved the view from our second floor apartment at 440 Jackson Avenue, she was scared of the height, having always lived on the ground floor. There were only two of us children at the time—I was eight years old, and my sister Millie was five years old. The park was a great place for us to play in and run around. There was no such thing as parents' supervision.

One of the first things my mother arranged in this new apartment was to have a clothesline strung on rollers between her bedroom window and a neighbor's in an adjoining building across the alley. Monday was wash day. She would wash clothes and sheets in the bathtub on a washboard and leave them in the tub all wrung out until I came home from school at lunchtime. Because she was fearful of heights, it was my job to hang the laundry and then take it off in early evening.

Afternoons after school, I was allowed to go out to play with my friends. There were no organized leagues for kid's sports. Instead there were gangs. I belonged to the Barracudas. My friend Salvatore (Sal) was our leader. His father arranged to have jackets made with the big fish showing its teeth imprinted on the back of the jacket. The price was $5. My father didn't allow me to get one. In the middle of the Depression, we couldn't afford it.

Since my sister Millie was always too young and I was the only boy, I had other chores. On Thursday afternoons, the fruit-and-vegetable man and then the ice man, both with horse-drawn wagons, would arrive in front of our apartment building. The women would shout their orders from their open windows. From our living room window, I lowered a basket tied to a rope with money in it. My mother's fruit-and-vegetable order and any change was placed in the basket, which I then pulled up. We had a dumbwaiter in the hallway of our apartment. I would lower it with the money for the ice wrapped in a piece of old newspaper. The block of ice was placed on the dumbwaiter. I would then pull up the ice and carry it to our icebox. I can't remember how old I was when the icebox was removed and replaced by a refrigerator manufactured by Frigidaire. One chore off the list. Our rent increased by $1 per month.

On Thursday nights I went with my mother to the butcher shop, which was always jammed and loud with women and chickens cackling. The chickens were in a back room, alive and in cages. The women, including my mother, groped the underside of the chickens to determine which one was fat enough. That was important, because in kosher homes chicken fat was a primary source for replacing butter when used with non-dairy foods. After she had selected a chicken, the butcher would take it out in the back, where someone else would slaughter it, drain the blood, and hand it to Mom. She would then sit down in the back of the shop and pluck the feathers. When she grew tired, I had to take over the plucking. It wasn't until after World War II that kosher chickens were available in a refrigerated showcase, already slaughtered and plucked.

Fridays were very busy cooking and housecleaning days. My mother was heavily involved in cooking meals for Friday night and Saturday. The Sabbath was meant as a day of rest, and she followed that rule. When I came home from school, my chores continued. First, I had to wash all the windows with a water-vinegar mixture and then use newspapers to rub and dry them. Because she was afraid of heights, I would stand on the windowsills to clean the top inside of the windows, then open them and sit on the windowsills to clean the outside. There was lots of soot from all the coal-burning furnaces. It took a few years to convince her not to hold on to my legs and that I would not fall down to the street below. After the windows, I washed the kitchen and bathroom floors. She then covered the kitchen floors with newspapers to keep them clean.

After my father was served his meal, my mother would take me to the movies. My father didn't care for movies and Millie was always too young, according to my mother. It didn't matter what was playing because every Friday night free dishes were given away — one dish for each person.

The Metropolitan Opera on the Air was on the radio Saturday afternoons at 1 p.m. during the winter and early spring. My mother insisted: "It's culture." I had to listen, but my sister did not. Girls just had to get married. I later found out that one of my mother's cousins was *wealthy* and had season tickets to the Met. She lived on Central Park West and took her son, who was my age, to Saturday afternoon performances. My mother was not to be outdone. So, I listened. After the opera season was over, I was allowed to go to the movies with my friends. Luckily the Saturday matinee movies were different from those on Friday nights. They were kid-friendly cartoons, cowboy serials, and cowboy double features.

Since my father worked six days a week, Sundays were togetherness days. It was the only day of the week that the family sat down to eat all meals together. I was also required to read the Sunday *New York Times*, especially the *Week In Review* and *Book Review* sections. This was meant to broaden my education. For late lunch (3 p.m.), which we called dinner, my mother always prepared a roast and a full meal. Different aunts, uncles, and cousins came every week. For dessert, the women tried to outdo each other with their pies or cakes. My boy cousins and I would go off to the park and play ball. The girls would play hopscotch or jump rope in front of the house. All of the relatives left by 6 p.m. It was then that the Marcus family gathered in the living room to listen to Jack Benny on the radio. Those were fun times.

The Sunday dinners ended in June 1936 when my mother gave birth to my sister Elaine. I was 14. It seemed that every evening while my mother was serving my father his supper, Elaine needed to be fed her bottle and rocked to sleep in her carriage. That turned out to be my job. No more evening walks with my father. Studying and homework became an early morning chore. I became an early riser and to this day, I still am.

Many things changed in May 1938 when I went to work on a farm, then attended City College in September. No more hanging out with the boys in the park or on the street and no more Friday night movies. My sister Millie was now required to do all the chores that once were mine. Liberation day.

Our family in 1939: Pop, Mom with Elaine, Millie, me

Walking To Junior High School

In snow and rain and heat and gloom, I would walk to Clark Jr. High School, 360 E. 145 St. from my apartment building, 440 Jackson Avenue. Clark was an all boys' school.

Classes started at 8:30 in the morning. Lunch hour was from 12 to 1. School ended at 4.

Leaving my apartment building, I turned left (south) and walked to St. Mary's Street. Along the way I picked up three buddies—Sam (Jewish), Louis and Salvatore (Italian). We would always go together, never alone. We then turned right on St. Mary's St. and turned right on St. Ann's Ave., walking alongside the park to 145th St. We then crossed St. Ann's Ave. When we arrived at the other side, we were in the Irish neighborhood, and we were usually met by a group of Irish kids also heading for school. Shouts of "dirty kikes" and "dirty wops" greeted us, and we answered with "dirty micks." Everyone walked in groups. If one of us walked alone, the taunting would end up with lots of shoving and running until we got to the schoolyard, which was three long blocks further to 3rd Ave. It took us about 20 minutes.

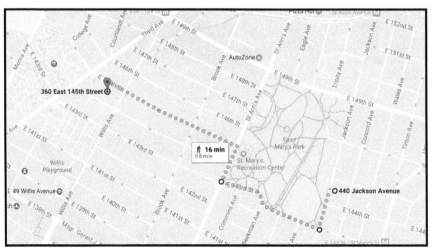

The route from our house to the junior high, as shown on a Google map

I remember once when I was late for school and my friends had already left, my mother walked with me to the schoolyard. That was embarrassing. However, no one messed with a mother from the Bronx. We would go home for lunch and then back to school again. When it rained, we ran. When it snowed, we had lots of fun tossing snowballs. Google says the distance from home to school is only 0.8 miles, but to a 10-year-old kid it was miles and miles.

Bronx Barber Stories

In the 1920s and 30s, our local barber had many other talents in addition to cutting hair. One of them was pulling teeth. When I was eight or nine, I had a badly infected back tooth. We couldn't afford a dentist. My father took me to the barbershop down the block from our apartment house in the Bronx. There was a sign in the corner of the window: "Teeth pulled — a quarter each." Pop sat me on the barber chair and held me down. The barber came over with a pair of pliers, which I hope he cleaned with alcohol, and pulled out my tooth. It hurt!

There was another sign in the window: "Cupping." Whenever I had a bad chest cold, my mother would go down to the barbershop and arrange for the barber to come to our apartment that night with a bag full of small glass cups and small candles. He would then light a candle and place a cup over it. When the cup was hot, he would pick it up and quickly place it on my back. It was a hot suction cup and apparently sucked the cold out of me. Depending on how bad my chest cold was, it took either three or six cups — a quarter for three cups.

If you were a teenager and had acne, for a quarter the barber would squeeze out the pimples with a "pimple popper." He would then clean your face with alcohol.

Every Friday night my father would give me a quarter with a piece of paper that had three numbers and his name on it. I was to take it to the barber and wait until I heard him call someone on the telephone in his back room. Then I could leave. I didn't realize that he was calling a bookie. All I knew was sometimes my father would come home and proudly give my mother $2, which he had picked up at the barbershop.

By the way, haircuts were a quarter.

The Mob—Irish, Italian, Jewish Mafia

Background: The capture of Whitey Bulger, the murderous boss of the Irish Mafia, in Boston in July 2011 brought to mind my earliest recollections of the Mob.

I was probably about six or seven years old and we still lived on 138th Street. It was either a Sunday or a holiday because my father was at home. Our family—my dad, mother, my sister Millie and I—were having lunch when we heard several sharp noises outside. My father shouted at us, "Get down on the floor, someone is shooting."

When things seemed to be quiet, he told us, "Get up, while I go to check things out."

He came back into our apartment and said that someone had been killed and the police and an ambulance were down the block. Several days later he read me a news item from *The Daily News*. A fellow had been picked up for the shooting. I don't remember his name, but I do remember that he was the president of the Pickle Union.

"A nice Jewish boy," my father said.

Several weeks later he told me that this person had been released from prison and that someone in Sing Sing Prison had confessed to the crime. The rumors on the street were that the Pickle Union guy's brother had something to do with the confession. The brother was a captain in the New York City Police Department. The other word on the street was that the Sing Sing prisoner's family would be well taken care of for the rest of their lives by the Mob, who ran the Pickle Union.

My uncle, Louis Altarescu, had a radio store down the block from our house on 138th Street. I would go and visit Friday afternoons after school and play on the street in front of the store with my cousins. Toward the end of the day a policeman would come into the store. I saw my uncle slip him an envelope, and then they shook hands. One day I was in the store when it happened. I overheard the policeman say, "It's too bad that the laundry store owner up the block stopped contributing. His windows were broken last Sunday."

When he left, my uncle looked at me and said, "Five dollars a week and the cops protect my store. They really belong to the Irish Mafia."

After we moved to Jackson Avenue, one of my two Italian friends was Salvatore. I can't remember his last name. He came from a large family, and they lived in one of the few single-family homes in the neighborhood. They were the only family in the neighborhood that owned an automobile. It seemed that his father was home most of the time, and there were lots of visitors arriving in cars. One day while we were walking to school, I asked Sal what his father did for a living.

"I don't know exactly." he said, "He works for the family."

Many years later, I realized he was with the Italian Mafia.

Looking back, I'm sure my parents were very worried about how the environment and my friends would affect my life. They tried to keep me off the streets as much as possible, especially on weekends and summers, and they also wanted me to be well informed and intelligent.

- I can distinctly remember my mother yelling at me from the kitchen window when I was playing in the alley with my friends from the gang (the Barracudas) to "Stay away from them bums."

- Every Saturday during the opera season, I had to listen to the radio broadcast of the Metropolitan Opera on the Air.

- When there was no opera, my other Italian friend Louis and I would be given subway fare and lunch by our mothers. We had strict instructions to go to one of the museums in Manhattan or to go to a double-feature movie. I'm sure the two mothers had a deal going.

- My father would buy the *New York Times* on Sundays. He insisted that I read *The Week in Review* and *The Book Review* sections.

- Sundays were family days—no problems on the street.

- During the summer, we would rent a bungalow in Rockaway Beach, which kept me off the streets. My mother had a coffee can in which she put leftover change so that we had money for this vacation.

- When I went off to high school and then to college, I'm sure my parents were greatly relieved.

In their wisdom, they kept me from getting involved with "them bums."

Thank you, Mom and Pop.

The Great Depression (1929-1940)

Background: It is March 2009, and this country and the world are experiencing a serious recession. I would like to share what it was like for me to live through the Great Depression between 1929 and 1940. The important thing is that with the help of the U.S. Government, we survived.

I was seven years old in 1929 when the Stock Market collapsed. My father came home from work with the evening paper and showed my mother the shocking headlines of the men who committed suicide by jumping out of the windows of their Wall Street offices. Fortunately, my father was not an investor. I remember my parents being very frightened as we listened to the news.

The financial crisis hit us about a year later when my father told us that he would only be working on weekends, earning $5 per day (his regular take home pay was $20 per week). The weekends-only work luckily lasted only a few months; however, he worked five days a week instead of six through many years of the Depression. He was an optician and had been working for the same optical store since he got out of the Army after World War I.

To make ends meet, he started selling eyeglasses from our apartment. His boss had told him he could use the equipment in the store to grind the lenses on his own time. On Saturdays, he would take me to work with him, where I stayed in the grinding shop in the back and watched how he ground lenses prescribed by an optometrist or ophthalmologist. It was undoubtedly there that my interest in optics developed.

We moved twice in order to take advantage of the free first- and last-month rent that landlords were offering with a signed one-year lease.

My grandfather, who was a tailor, lost his job in the garment industry. He, my grandmother and their four youngest children lived on the lower east side of Manhattan. On Thursdays, my mother would buy a shopping bag full of groceries and a chicken, and give me ten cents to ride the subway after school to deliver the food. I was ten years old at the time, and my parents had no qualms about my traveling alone. It took about 30 minutes on the subway with a 15-minute walk at each end. That went on until my grandfather went back to work many years later.

In 1932, my father, who had been wounded in World War I and received a Purple Heart Medal, joined a new organization—the Military Order of the Purple Heart. Early that year there had been a Veteran's March to Washington, D.C., to call on the President and Congress to provide the bonus to veterans that had been promised them for their service during the War. Pop left on a bus with other Purple Heart veterans to join the marchers in Washington, D.C., but thankfully, he returned the next day. The bus had been turned back and couldn't enter Washington. We found out later that President Hoover had ordered General MacArthur to evict the veterans from Washington. The General ordered the troops to fire on the veterans, whom he called Communists. Many were killed. The bonus was finally passed and paid in 1936. My sister Elaine, dubbed the bonus baby, was born in June, and my father received his bonus in July in the form of United States Bonds. They were worth $600. That was a fortune.

Saving pennies was a must. During the winter months, we had a metal window box outside our kitchen window, as did everyone else in the apartment building. When winter came, my mother would not buy ice for the icebox. All the perishables were stored in the window box. Luckily our kitchen faced a shady alley. We finally were offered a refrigerator for $1 extra rent per month, which we took, and the icebox and window box disappeared. At 10 p.m., the building furnace was shut down to conserve coal and turned on again at 7 a.m. It took a long time for heat to get to our second-floor apartment.

In the mid 1930s, after the election of Franklin D. Roosevelt as President, the Works Progress Administration (WPA) and the Civilian Conservation Corps (CCC) were created. Those two agencies proved to be lifesavers, putting millions of people to work building roads, bridges, public buildings and parks. The WPA also provided food for school children, work for artists and musicians and many more. The CCC provided work for men in natural resource conservation. My closest cousin, Nat Altarescu, joined the Corps in 1938 and went to North Dakota. I was very envious. One had to be eighteen. I was sixteen.

The WPA was also a big help for my family. There was a grant to public colleges to hire students, whose families had a minimum or no income, to perform various jobs on campus. After I started City College in September 1938, I qualified and was given a job to be an assistant in the Physics Department labs. I worked ten hours a week after my classes, and earned $10 per month. The minimum wage of 25 cents per hour had just been instituted. I would hand over my check to my mother, who was the family banker.

Speaking of banks, people had zero confidence in putting what little money they had into savings accounts because of all the bank failures. To encourage people to open an account of $100 or more, banks would give a gift of some sort. My father was given our first toaster from the Bowery Savings Bank. A few months later, he took his money out and put it into another bank, which provided a small radio. That radio, cherished by my mother, ended up in our kitchen where she would prepare meals and listen to her soap operas at the same time. Stella Dallas and Molly Goldberg were her favorites.

Going to the movies was a must, especially to see lavish musicals. It provided an escape from the real world. My mother would take me with her on Friday nights so we could get two pieces of china. She ended up with at least a complete service for eight. Movies were 25 cents for adults and 15 cents for kids.

In 1940, with the war in Europe going badly for France and England, President Roosevelt instituted the Lend-Lease program to supply England with all types of war materiel. The Depression started easing. My father began to work a full week, and I lost my WPA job. I graduated from City College in 1942 and went into the Army. Most of what I earned there went home to my family. To my good fortune, my mother saved it for me. The beginning of World War II marked the beginning of the end of the Depression.

Halloween in The Bronx In The 1930s

Times and customs have indeed changed.

"Trick or Treat?" I never heard of it until the late 1950s when my kids were old enough to be taken from one neighbor's house to another to load up on candy.

Halloween in the South Bronx when I was growing up was a time for big bonfires and lots of rowdiness. Until I was thirteen years old, I wasn't permitted to go out of the house on Halloween. At thirteen, I became a sworn member of our local gang, the *Barracudas*. The neighborhood gang was made up of Jewish and Italian kids. My last Halloween out on the streets was on October 31, 1935. The Barracudas held a meeting in our usual alleyway the previous day, when our gang leader gave us our instructions for Halloween.

First, we were to go around the neighborhood, collect all the pieces of wood we could find, and stack them in the middle of an empty lot nearby. One of the places we went was a garage where horse-drawn wagons were kept. We found a broken-down wagon that one of the workers let us pull away. We broke it up into small pieces and stacked the wood up high. The idea was that the gang that had the biggest fire in the surrounding neighborhoods was *Number One*.

In order to protect our large stack of wood so that other kids wouldn't take any away for their fires, we were each assigned a time to stand guard until the fire was lit at night. The way we protected our stack was to form a circle around it. Each of us carried a weapon, a silk stocking that we somehow obtained from our mothers. The stockings were filled with sand, talcum powder and small rocks, which we used to swing at any kid who came too close.

About 8 p.m., our gang leader started the fire. We ran around it and cheered. It was a big one. Then someone screamed that a nearby building had caught fire. Apparently, some sparks had landed on a roof. Soon we heard the sirens of the fire engines and the police. We all ran in different directions. I ran home, as did most of my friends. My mother and father hid me in the bedroom and looked out the window to make sure no one was coming after me. The fire was put out rapidly, causing very little damage; but that was the end of my short membership in the Barracudas and my last Halloween until I walked Deborah and Victor from house to house collecting candy.

Bronx Boy On A Farm

It was January 1938. I was 15 years old and had completed all my courses at James Monroe High School. I was really young. In elementary school, I had skipped grades because I was at the top of my class. Also, my mother had arranged for me to start school in kindergarten six months early. I had been accepted to attend City College of New York for the spring semester.

My father said, "No, you're too young to start college. I want you to work on a farm until September. It will help you grow up. I worked on my father's farm and in his bakery in Romania until I was 16, then came to this country. I was grown up."

Growing up in an apartment house, my biggest job had been washing the windows and floors on Friday afternoons and taking care of my baby sister, Elaine.

One of my cousins, Max Altarescu, spent each summer in the Catskill Mountains with his family as paying guests on Mr. Prager's dairy farm in Accord, New York. Mr. Prager had cottages on his farm that he rented out in the summer. He hired help those months to do a variety of chores. My cousin and my father spoke to Mr. Prager, who said that I could start working on his farm on May 15 (my birthday is May 14). Wages were $20 per month, payable at the end of the summer, plus room and board. With trepidation, I boarded a bus to Accord. Mr. Prager met me at the bus station. He appeared tough and barely spoke to me. We arrived at his farm at about suppertime. He introduced me to his wife, who was much friendlier. They seemed old, probably in their 30s.

They had two dozen cows and lots of chickens. The workday started at 5 a.m. First, the cows had to be milked, and then the barn had to be quickly cleaned up. After that, we fed the chickens and collected all the eggs. There were also two horses that had to be cared for.

I was told I was to be their only farmhand and that I would be sleeping in the attic. Mr. Prager then recited my schedule:

"Each day starts at 5 in the morning with a cup of coffee or buttermilk.

"The cows get herded into stalls in the barn, are fed, milked and then driven out of the barn.

"The barn gets cleaned up and washed down.

"The cows get driven out to the pasture.

"The chickens get fed and then the coops get cleaned.

"The eggs get collected and then get candled.

"Feed the two horses and clean their stalls.

"Breakfast is at 8:30.

"After breakfast, we go out into the field and you will be given chores to do.

"Lunch is from 12 to 12:30 in the field. Mrs. Prager will bring out the lunch.

"At 4, round up the cows, drive them into the barn, milk them and then clean the barn.

"Feed the chickens, then clean the coops.

"Then wash down and feed the horses."

I couldn't imagine doing all that. The schedule seemed overwhelming. Mr. Prager said we would work together in the beginning. When the summer vacationers arrived in late June, all from

New York City, I would be expected to do much of the early morning work on my own, since he would be helping his wife prepare the meals for them. I thus discovered what the American plan was: he rented out eight cottages and provided meals to the guests.

What I can vividly remember about the early days was the difficulty I had learning how to milk the cows and how much my thumbs hurt. In the beginning, Mr. Prager would make sure that I milked each cow dry. If I didn't, he would yell at me. When I was finished with each cow, I would try to get her out of the barn in a hurry so that I would have a minimum to clean. I remember having a cup available to taste the warm, rich milk that had filled the pail—what a wonderful flavor. We had no days off. The cows had to be milked twice a day, and the chickens fed and eggs collected. On Sundays, we didn't work in the fields. As time went by, I felt that I was developing muscles and getting stronger and better at my job.

Being part of the next event was the most amazing experience I had there. One day one of the cows didn't come in with the herd when I rounded them up. Mr. Prager and I went looking for the cow and found her lying down. "Help me," he said, "She's giving birth and having trouble."

He had me sit on the cow while he pulled out the calf feet first. He cut the umbilical cord and had me clean up the calf and get it on its feet. He told me that the calf would stay with its mother overnight in the field. The following morning we separated them. I received yet another responsibility—to feed the calf until it was on its own. Meanwhile, the mother would be a great source of milk.

Late in July, the hay had to be cut, dried, loaded onto a wagon and then tossed into the hayloft; nothing was motorized. There were two horses, plus Mr. Prager and me. The horses pulled the cutting machine and then the raking apparatus, and I followed behind them holding the reins. Mr. Prager was the guide in front. Before we started working, I was told to feed the horses lots of garlic, which Mr. Prager said helped them work hard in the heat. When the horses pulled a loaded wagon of hay slowly up a hill on the way to the barn, they would expel gas right at me, which smelled like garlic plus manure. To this day whenever I smell a heavy scent of garlic, I can see the horses' tails rise. The toughest job then began, which was tossing the hay into the loft of the barn, and then stacking it. The hayloft was very hot.

One Sunday afternoon, Mr. Prager said that we would load two cows onto his truck and take them to get "fixed." We drove several miles to another farm. I was introduced to the farmer who said, "Boy, unload those cows and lead each one into a separate corral." I soon discovered that the farmer had let loose a bull in each corral, and I would get a lesson on the birds and the bees.

It was the last week in August and my time was up. I was ready to go home and start school. Mrs. Prager prepared a great farewell dinner party, which included all the vacationers. The next morning, Mr. Prager paid me my $70 in wages, then took me to the bus stop and said, "Come back next year."

I didn't go back.

Tammany Hall—New York City

Background: On December 9, 2008, the Governor of Illinois was indicted for corruption and soliciting bribes. This brought to mind my association with Tammany Hall when I was in my teens.

Tammany Hall was the Democratic Party political machine in New York City. It was formed in the late 1700s with the large influx of Irish Immigrants into this country. It helped them get jobs in exchange for a fee and also for voting for the Democratic Party. It grew considerably as the large numbers of Italians and Jews arrived in New York from Europe. All city employees were required to become members and pay monthly dues according to their job. Non-employees paid $2 a month, for which they could go the Hall and ask for a favor. It was like belonging to a large and powerful union.

My father joined Tammany Hall when he became a citizen after serving in World War I. Tammany Hall helped him by arranging an apprenticeship to become an optician and then helping him to get a job. He never told me how much that cost him, but it did make him a lifelong Democrat.

When I was 13, he took me to the Tammany Hall offices in Manhattan to find a job. I was hired to stand at a table outside the Federal Courthouse with several other kids when immigrants left after being sworn in as citizens. I was to tell them that they now must sign up as members of the Democratic Party. For each new member, I would receive 25 cents. The new citizen would receive $1 with the understanding that they would vote for the party candidates at all elections. I thought that was great and was assured I could always count on Tammany Hall if I needed help.

When I was 16, the country was still in a deep economic depression. Jobs were hard to come by. My father arranged for me to have an interview at Tammany Hall for a job. He told me that Tammany Hall controlled part-time jobs for the U.S. Postal

Service, particularly during the Christmas holidays. The interview went smoothly after I enthusiastically agreed to their terms: when I received my pay check from the government, I would bring it in to be cashed at Tammany Hall, I would then keep 90 percent and they would get the rest.

I was assigned to the main Post Office on 32nd St. for a ten-day job. The Post Office worked around the clock. Tammany Hall also arranged hotel rooms at the Hotel New Yorker for "their employees." We were encouraged to work 16 to 20 hours a day and sleep for several hours at the hotel. I would stay in Manhattan for the entire holiday period. The job was a great help when our family needed money for food and rent. I worked at the Post Office during the holidays when I was 16, 17, and 18.

Several years later, the Mayor of New York, Jimmy Walker, suddenly left for Europe, where he stayed to avoid prosecution for bribery, corruption and extortion. That was the beginning of the end for Tammany Hall. We couldn't understand what the problem was since they did so much good for the poor of the City.

I'm Colorblind

In writing my memoirs, every once in a while a colorblind story appears.

Therefore, I decided to record all my colorblind stories in one place. I'm red-green colorblind, as are about 20 percent of males. The genes are handed down from mother to son and are passed down through daughters.

In 1928, when I was six years old and in first grade, the teacher sent me home with a note on a drawing of a cow, which I was to color with specifically identified colors. The note read, "Please teach your son colors." I apparently mixed them up. My mother told me she would help me with colors just the way my grandmother helped my grandfather. She said that he never learned colors even though he was a tailor. My grandmother picked out the correct threads for him when he sewed a garment. My grandfather was colorblind. As I was growing up, coloring was not my thing.

I graduated from City College of New York in June 1942 with a BS in Physics and had also completed four years of ROTC training. This was six months after Pearl Harbor and the United States was heavily involved in World War II. About a week after graduation, I received orders to report to Maxwell Field, Alabama, to receive my commission in the Army Air Corps as an engineering officer. After a long train ride and overnight in the barracks, all of us cadets reported for physical exams. I flunked the color test and was sent back home and told to register for the draft.

I was drafted into the infantry in September 1943 and was sent to Camp Wheeler, Georgia, for basic training. Since I was a physicist, I was sent to an antiaircraft outfit to be trained as a radar repairman. That didn't last long when instructors discovered that I was colorblind and couldn't tell one resistor from another. I was transferred to the brigade headquarters, became an assistant to the intelligence officer and quickly rose in ranks to staff sergeant.

In January 1944, the Armored Force sent a message to all military bases that it was opening a special Officer Candidate class for 100 qualified staff sergeants and above. I applied and was accepted, pending a physical exam. I knew all the medics, and when I came to the medic who had a sign on his desk reading Color Test, I looked at him and asked if $20 would pass me. He said "Yes." I was soon off to Fort Knox, Kentucky, which was the headquarters of the Armored Force. Cadets in training were very supportive of one another since we all wanted to graduate. My closest buddies knew that I was colorblind. One night when we were out on maneuvers, the officer in charge shot a flare into the air, swung around, pointed at me and shouted "You, Cadet, what color is that?" My buddies whispered "Green." "Green," I shouted. The officer said, "That's right. In my last class I shot a flare into the air; the cadet was colorblind, and we threw him out of school."

After graduation, I was sent to the Military Intelligence Center in Fort Ritchie, Maryland. At the first meeting of the newly arrived officers, the officer giving the briefing asked for a show of hands of any officer who was colorblind. Nobody raised his hand. He went on to say that our commissions would not be taken away, but that the Army was in need of photo interpreters who could look at color aerial photos and detect camouflage. I raised my hand and was sent to take photo interpretation training. I could not see through camouflage.

With the invasion of the European continent, Counter Intelligence Officers were sorely required as our armies headed for Germany. Because of my German language training in college, I soon became a Counter Intelligence Officer and went overseas. Nobody really cared whether I was colorblind or not.

In October 1946, after I was released from active duty, Molly and I were married and living in Brooklyn. I bought my first car and needed a driver's license. I went to the local DMV office and applied. I was taken out for a driver's test. The Brooklyn street traffic wasn't too bad. After I stopped for a red light (I can tell the red from green, the red is darker and on top, and the green looks white), I was told to pull over to the side. The officer then took out a bunch of colored yarns and asked me to identify them. I failed miserably. "Well," he said, "you know when to stop for a red light, but I can't pass you." Since it worked once before, I took out a $20 bill and put it on the dashboard. "I made a mistake," he said "You pass."

In 1953 we were living in Eatontown, New Jersey. My daughter Deborah was 5 years old, and my son Victor was 3. Molly and I always liked to take the kids to Asbury Park, where there was a boardwalk and all kinds of fun rides. One day, as we were driving down the main street, I noticed that a new traffic lighting system had been installed.

There were now three lights, and I could barely tell the difference between the amber and red light. To add insult to injury, the lights were reversed at every major intersection, that is, changed from red on top and green on the bottom to green on top and red on bottom. Molly kept us safe and would tell me when to stop and when to go. The kids then picked it up and whenever we went for a drive, they would yell, "Stop Daddy!" or "Go Daddy!"

When Ruthie and I were married in March 1993, she discovered that I was colorblind. She inherited the responsibility of Molly and the kids. Every time I am driving on city streets with traffic lights, I can be assured I'll hear her say, "Stop Bernie," or "Go Bernie."

I'm in the Army Now

"You're in the Infantry!"

I graduated from CCNY (City College of New York) in June 1942 with a BS in physics and had also completed four years of training in the ROTC. I was 20 years old. After graduation I was sent to Maxwell Field in Alabama to receive my commission as a second lieutenant in the Army Air Corps. Because I flunked the color-blindness test, I was sent back to New York to register for the draft. Drafted into the Army in September 1943, I was sent to Camp Upton, Long Island. This is where my dad was sent when he enlisted in the Army in 1917. The advice he gave me as I left home was "Don't volunteer for anything and always play *Mickey the Dunce* until you know what's going on."

Me in my ROTC uniform when I graduated

After we arrived at the camp at about noon, we were lined up at the infirmary. As we entered, we were told to take our clothes off, except for our shorts, for an inspection by the medics. They inspected us all over, and I mean inside and out. At the end, each of us was quietly asked by an Army doctor with a wink, "Do you like boys or girls?" At the time I had no idea what happened to those who said they liked boys. Later I found out that they were dishonorably discharged. Then there were the shots. I had no idea what they were all for, except that it seemed endless as we walked along a ramp and received shots in both arms. I remember that several guys passed out when they saw the needle pointed at them.

Then we filed by a counter, and the GIs behind it were shouting, "What size jacket? What size pants? What size shirt? What size cap? What size shoes?"

One at a time the items were "gently" thrown at us, including underwear, a tie, fatigues, toiletries, and finally a backpack. We were told to dress up in our new underwear and fatigues, fill up our backpack, carry our uniform, and line up outside.

A sergeant then walked us to our barracks and assigned us bunks. Footlockers were located in front of the bunks, and we were told, not politely, to organize everything neatly in it. On top of each bunk was a sheet, a pillow, pillowcase, and a blanket. The sergeant bellowing instructions showed us how to make our beds and then came to inspect. If he didn't like what he saw, he tore it apart and watched as we made our bed again. There were about 50 of us in our barracks, in upper and lower bunks.

After he appeared to be satisfied, he ordered us out of the barracks and lined us up. Then he shouted, "Anyone here go to college, one step forward."

Three of us stepped forward.

"You three, you are on latrine duty until further notice. Cleaning time is 6:30 in the morning after everyone has showered, and 9:30 at night before lights out."

"Further notice" turned out to be the entire week I spent at Camp Upton. So much for a college education and not following my dad's advice.

On latrine duty

He informed us that we would be leaving Camp Upton within a week for a basic training camp. We were told to go into the barracks and relax until chow time, which was 5:30.

My first meal in the army was a new experience. We lined up in the mess hall to get our food. We each had our food, including dessert, dumped onto our tray. I sat down and looked at the food, and there to my surprise was a very large lamb chop.

I cut a piece and bit into it. It was tough. I told the fellow next to me: "This is the largest and toughest lamb chop I have ever had."

"What, are you crazy?" he said "This is a pork chop."

I couldn't eat it. Pork was a forbidden food. Later that week at the Friday night service, an Army rabbi told a group of us Jewish GIs that we could eat pork and that God would forgive us. He didn't want us to starve. Breakfast the next morning was another new experience. We were served a slice of toast, and then a thick, creamy meat topping was tossed on top of it.

"What is this?" I asked the GI serving. "SOS" was his reply, "You'll like it."

I did.

After a week of calisthenics, learning to march, and cleaning up the camp, we were given our orders to leave—destination unknown. One morning after breakfast we were told to put on our uniforms, pack our gear into our backpacks, remove our sheets, pillowcases, and blankets and stack them in barrels, then assemble outside. The sergeant told us to board the waiting trucks.

We were on our way to our basic training camp, where they would make good soldiers out of us. The trucks took us to a station, where we boarded a waiting train. Each car had a sergeant in charge, who bellowed instructions. We were told that we would be traveling for several days, and he had no idea where we were going. We would be stopping along the way for chow at train stations, where the Army would be setting up mess tents. He showed us how to prepare bunks for sleep at night. There were no sheets, pillows or blankets. We were roughing it. We traveled five or six days, zigzagging through Pennsylvania, Ohio, Kentucky, on and on. One of the guys said he heard that we were doing this because we didn't want the Germans to know where we were heading. We also needed the Germans to think we were assembling a very large army.

Finally, we arrived at Macon, Georgia, and were told we were at our destination. It was about 10 p.m. There were a couple of hundred of us who then climbed aboard trucks for a ride to an Army camp with a large banner at the gate:

Welcome to Camp Wheeler

We were lined up in front of a large platform, which was all lit up, when somebody shouted "Attention!" through a loud-speaker.

Then in a very loud, thick southern accent we heard, "At ease, I'm Colonel Jones, you're in the infantry. We're going to teach you how to shoot, how to kill, and how to protect yourself in close combat with a bayonet. You will be crawling on your bellies under barbed wire while we shoot live ammunition above you from a machine gun. As an infantryman, you will be going on 25-mile hikes with a 60-pound load in your backpack. We'll teach you how to march in a parade and take orders without questions. This training will take eight weeks, and then you will go overseas. And don't any of you New York Yankees think you can transfer out of the infantry."

The fellow standing next to me whispered: "My mother-in-law wished this on me."

In and Out of the Infantry

After the great welcome into the infantry, we were shown to our barracks and assigned bunks. A giant with a loud southern accent told us to put our clothes in our footlockers, shower, and go to bed. The next day would be our first full day as real soldiers. It must have been after midnight when lights were turned out and I fell asleep.

No sooner did I close my eyes when I heard a bugle and then a voice shouting, "Up and at it, you have ten minutes to put on your fatigues and line up outside for roll call."

I looked at my watch; it was 5:30 in the morning. We did as we were told and lined up in the dark.

The voice announced, "I'm Sergeant Smith, in charge of making soldiers out of you. The first thing we're going to do is have a little jog around the base for half an hour to warm us up for the rest of the day. After that we line up for chow at the mess hall."

I don't think any of us ever ran that far before breakfast or ran for exercise. If we ran, it was either after or away from somebody. It was exhausting.

During the remainder of the week, we were issued M1 rifles with bayonets. We had target practice shooting at figures of men at various distances and charging at stuffed figures with our bayonets. Then began the 25-mile hikes with a 60-pound pack on our backs while carrying a rifle and bayonet. I barely weighed 130 pounds.

We were permitted to make telephone calls during the weekend. The lines at the telephone booths were long, and of course we all called collect. When my turn came at the end of the first week, I told my parents where I was and asked my dad if he knew anyone who could get me into a technical branch of the service. He wasn't too sympathetic since he had served in the infantry in World War I. He said he would think about it.

When I called him the following week, he said that the only idea he had was for me to write to our congressman, Representative Samuel Dickstein, who was also a member of Tammany Hall, and explain the situation. He gave me his New York address. I wrote him a letter emphasizing that the Army was sending GIs to college to study physics and engineering so that they can serve in the technical branches of the Army. Could he arrange for me to be transferred?

The first month went by:

- More 25-mile hikes. We often sang *Marching through Georgia,* as farmers would hurl potatoes at us yelling "Goddamn Yankees."

- The infamous crawling on our bellies under barbed wire while a machine gun was firing above us.

- Training with carbine, submachine gun and rifle. By the way, I received an expert marksman medal for all three.

Expert Marksman Medal

We were finally given a full weekend off. Three of us decided to go to Savannah, approximately 125 miles east of Camp Wheeler. We took a public bus and sat in the only available seats up front. A few miles down the road, the bus stopped to pick up a young pregnant black woman who held a small child in her arms. Two of us stood up to give the woman and her child our seats. The bus driver, very livid and very vocal, told the woman to stand up and go to the back of the bus. We objected because there were no empty seats back there. He ran off the bus cursing at us. A few minutes later a police officer stepped onto the bus with a gun drawn.

He ordered the woman to get up and go to the back of the bus. He then turned to us and said, "You goddamn Yankees get off this bus and go back to where you came from."

He told us to get on the next bus back to camp and never to come through his town again. We were shocked and frightened. We waited several hours for the next bus back to camp. There was a coffee shop nearby, but the waitress refused to serve us Yankees. The word had spread around.

It was about the seventh week of basic training when our sergeant came up to me in the barracks and told me to report to the company commander in his office. "Why?" I asked. He didn't know.

The captain told me to sit down. He handed me a folder, which he said contained my service records to date and a document that he said contained orders for me to report as soon as possible to an antiaircraft brigade at Camp Stewart. The camp was about 60 miles east on the way to Savannah. He said that he had no idea how this had happened, but I would be given a jeep ride to the camp.

I was out of the Infantry! But why an antiaircraft brigade?

An Antiaircraft Brigade

My jeep driver, a fellow private, drove me into Camp Stewart. The MP at the gate directed us to brigade headquarters. When he dropped me off, I remember him saying goodbye, and "You're a lucky son-of-a-bitch."

I reported to the sergeant in the headquarters building and handed him my orders.

"We've been expecting you," he said. "You will be reporting to Major Johnson, the Brigade S-2" (S-2 is the intelligence officer).

"I thought I'd be assigned to something technical," I said.

"This is the only vacancy we have, and you're a college graduate. Major Johnson's office is down the hall. Take your orders and report to him." I did.

Major Johnson told me that I would be his assistant, particularly during several war games that were planned before the brigade went overseas. He gave me some books to study on the history, make-up and tactics of an antiaircraft artillery unit. He told me to report back to the master sergeant, who would show me to my quarters and the mess hall and would advise me of my additional duties. I did.

The master sergeant was very nice and appeared happy to see me.

"Since you are now the second private in brigade headquarters," he said, "you will be sharing the following duties with your new-found buddy: tidying up and cleaning the officer's quarters after chow in the morning before reporting to Major Johnson and reporting to the mess sergeant three times a week before chow in the morning."

To say the least, I was one busy guy. Major Johnson spent a great deal of time with me visiting the antiaircraft batteries under the brigade, which included radar and searchlight units. He told me that in two months there would be one week of war games, which would be observed by a team from Washington. He said that this was very important for him because his performance would determine whether he would be promoted to lieutenant colonel, and all the help I could give him would be greatly appreciated. Two months went by very quickly. The Pentagon inspection team arrived. One of the officers was a lieutenant colonel from Army Intelligence. We bivouacked way out in nowhere Georgia, and the games began.

One event above all sticks in my mind. It was our first night in the field. The master sergeant showed me the general's tent and told me that I was to dig a slit trench for the general next to his tent. I was given a shovel and told to "get with it."

I remember digging and cursing out loud and then hearing a voice. "Have you got a problem, private?" It was the general. "No sir," I said. I kept digging.

At the end of the week there was a critique. Major Johnson came and told me that he had received a superior rating. He said that the colonel from the Pentagon had also been impressed with my performance and wanted to meet me. The colonel said that he was aware of the assistance I gave Major Johnson during the exercises. The S-2 was responsible to identify friend or foe aircraft, and he observed that the major and I made an exceptional team. The colonel gave me his card and said that if I was ever a cadet in an officer candidate school I was to write to him before graduation and tell him which school I was attending. He said that I would be a good intelligence service officer. We went back to Camp Stewart.

Several weeks later, Major Johnson told me that he had been promoted to Lieutenant Colonel and he and the General had agreed that I should immediately be promoted to staff sergeant. He handed me my stripes. No more KP. No more cleaning up for the officers. No more digging. The brigade master sergeant congratulated me and told me that I would be sent to radar school and then to antiaircraft training for a total of three months, and then back to the brigade from which I would probably be sent overseas. About a week later I was sent to school in Fort Bragg, North Carolina.

Training completed, I returned to Camp Stewart to again serve as an aide to Colonel Johnson. A few weeks later, he told me that the brigade had received a memo from the Armored Force headquarters at Fort Knox, Kentucky, regarding its Officer Candidate School (OCS). They wanted to sign up 100 first three graders (staff sergeant through master sergeant) from around the country for a 16-week crash course to commission second lieutenants. He and the General would like to recommend me, if I agreed. Copies of the recommended sergeant's service records and current physical exam were to be sent to the Armored Force OCS, and those selected would receive orders to report to Fort Knox.

What an opportunity! I agreed. But what a dilemma. Suppose they found out that I was colorblind—no OCS. I checked out the infirmary to find out when a buddy of mine would be on duty and in charge of leading me through the exam. When we came to the colorblind test, he said, "You're not colorblind, Bernie, are you?" "No way," I said. He didn't give me the test. I gave him a $20 bill. Late in August 1944, Colonel Johnson called me into his office and congratulated me. I had been selected and had orders to report to Fort Knox in two weeks. He granted me a week's leave to go home to visit my family. It sure was a nicer and certainly faster trip on a passenger train rather than the troop train we took to Georgia. It had been about a year, and it was great seeing my family. My dad was very proud of his son being a staff sergeant who was about to go to OCS. I returned on a train to Louisville and then a bus to Fort Knox, where I reported to OCS headquarters.

Armored Force OCS

At OCS headquarters, I was given directions to the barracks for cadets, which turned out to be a very large brick building. I checked in and was directed to an auditorium, where I waited for further instructions. After our full complement of 100 cadets had arrived, a colonel came to the stage and addressed the group. He told us the following:

- The course would take 16 weeks, following which we would be commissioned second lieutenants in the Armored Force.

- We would be divided up into units of 20 cadets. An Armored Force lieutenant would be in charge of each unit, and each unit would have its own barracks room.

- After the meeting we would be provided with new fatigues and uniforms that would identify us as cadets.

- The officer in charge of each unit would hand us a weekly schedule in our barracks room. Activities would include calisthenics, classroom instruction and field work. Our schedule would cover Mondays through noon on Saturdays. During that period we were required to run, not walk, every time we left a building and were outdoors. A violation could terminate our cadet status.

- Should any of us report to the infirmary for any illness, our cadet status would also be terminated, and we would be returned to our previous outfit.

- We would not be allowed to leave the base until commissioned as officers.

Each of our names was called, and we were assigned to an officer. After dismissal we followed our lieutenant to collect our new clothes and go on to our barracks. He gave us the word that this was going to be a tough 16 weeks. It was his job to find out if at least one of us didn't have what it takes to make an armored force officer. That cadet would be dismissed from OCS. He also told us that we would be shown how to make our bunks, and during his morning inspection he would toss a coin on the sheet of one of our bunks. It had better bounce. If not, the cadet would be charged with a demerit. Ten demerits would cause dismissal. On and on!!

He then outlined what a typical day would be:

- At 5:30 a.m., a bugle would sound a wake-up call. We were to get into our fatigues and assemble outside for roll call, followed by a 30-minute run around the track, then 30 minutes of calisthenics—rain, snow or shine.

- Showers next, and then assemble outside at 7 and run in formation to the mess hall for breakfast.

- We could return individually to our barracks to clean up before assembling at 8:45 to run to our first class.

- Classes ended at 4:30. We would then run in formation to our barracks, change out of our fatigues into our uniforms, and get ready for the lowering of the flag.

- After that, we would run in formation to the mess hall for dinner. Following dinner, we were on our own. We were expected, however to read all the material handed out in class.

- Lights out at 10.

After dinner that night, the 20 of us sat in our barracks room and pledged to help each other in the event of a problem, vowing to finish the course successfully. The senior sergeant asked each of us if we were aware of any problems we had that might get us kicked out. One cadet said that he had a very bad cold and was feeling sick. The sergeant said that he had been stationed at Fort Knox and knew a key medic on whom he could call, who would come over to the barracks to help anyone in need without snitching. He would have the medic come to the barracks and help the ailing cadet. I told the group that I was red-green colorblind, which I was sure would not be a problem in fulfilling my duties. The sergeant assured me they would back me up.

The next several months were really tough. There are a few events that are stuck in my mind:

- It was near the end of February, and being in Kentucky meant lots of rain and wet snow. I had come down with a very bad cold. One night after dinner, I felt very hot and realized that I had a fever. I told the sergeant and he had his medic buddy come over. He gave me aspirin to take every few hours and also gave me a shot of penicillin. By the next morning my fever was gone and I felt better.

- We had tank-driving exercises once a week. Each cadet had to drive several miles over rough terrain in M-4 Sherman tanks. On this memorable day it was snowing pretty hard, and we were driving over a rocky hill when we heard a loud clanging noise, and the tank stopped. Luckily, I was not the tank driver. The four of us climbed out and, to our chagrin, saw that we had a broken tread. The rules of the game were that the crew had to repair the damage. There were sledgehammers and repair parts on board. We tried and tried, but we could not repair it. I could barely lift and wield the sledgehammer. We finally radioed for help, which arrived eight hours later. The repair crew got us going. We, the crew, earned one demerit each.

- One night during maneuvers, our group of 20 was sitting on the ground alongside our tanks attending a lecture by our lieutenant on night-time fighting. All of a sudden he shot a flare up into the sky and turned around, pointed at me and shouted, "You, Cadet, what color is that?" The fellow sitting next to me whispered, "Green." "Green," I said. The lieutenant said, "Correct. In the last OCS class I did the same thing, the cadet said white, and he was flunked out."

As you recall, I had been given a card from an Army Intelligence colonel when I was in the antiaircraft brigade. He told me to write to him if I were to be in OCS and to let him know in what school I was. About a month before graduation, I decided to write to him.

One morning, two weeks later, before classes began, a major came to our barracks and asked me to accompany him to the Commanding General's office. I was sure that they found out about my color blindness and that I would be on my way back to Georgia. But when I got there the general said, "Cadet Marcus, I don't know what this is all about. I have received orders from the Pentagon to swear you in as a second lieutenant immediately, and to give you this envelope, which is marked *Secret for Lt. Marcus Only.*" The General swore me in, gave me the second lieutenant's bars, Armored Force insignias and an order to a tailor in Louisville to outfit me in a uniform immediately.

He congratulated me and told me to read my orders in private. I was not to tell my fellow cadets any of this and to leave from the tailor's shop for my destination. The orders instructed me to take a train to Baltimore where I was to go into a room marked Fort Ritchie. I would then be transported to the Military Intelligence Center, Fort Ritchie, Maryland. The name of the fort and its location were to be kept secret from family and friends.

Fort Ritchie

I took a train from Lexington, Kentucky, to Baltimore. I followed the instructions I was given and entered a room at the train station marked Fort Ritchie. It was an office occupied by a sergeant behind a desk and several armed MPs. The sergeant asked for my name, serial number and ID. He looked at a sheet he had with names, found mine, and told me that a bus would take me to Fort Ritchie. I was asked to step into another room and wait. There were about 20 officers and enlisted men in the room, many of them speaking foreign languages. I recognized German and Italian. I found a seat and waited. About an hour later, an enlisted man led us out of the train station to a bus. We were told that our trip would take about two hours.

About a half hour after we left Baltimore, we started driving up a winding road in the Catoctin Mountains. When we reached the top, we took a right turn. Marines guarded and blocked the road to the left. I later found out that the road led to "Shangri-La," President Roosevelt's mountain retreat, now known as Camp David. A short distance later we came to another blocked road. This time there were Army MPs. We passed through the gates and drove into Fort Ritchie. The following photos were obtained on Google.

Fort Ritchie gates

Lake within Fort Ritchie

What a sight! There was a lake in the middle of the Fort. From the streets of the Bronx to this. I'd never seen anything so beautiful. All the officers' quarters and enlisted barracks were made of bricks. We were taken to the administration building, where we checked in, were assigned quarters, and were given a map of the fort, as well as instructions for the next day. The officer's quarters were individual two-story buildings. Each officer had his own bedroom and bath. I thought to myself, *Is there a war going on?*

Officers quarters at Fort Ritchie

The officer's mess was comfortable and the food was outstanding. I found out that the cooks were all draftees from some of the most outstanding restaurants in New York City. They were sent to Fort Ritchie because they were fluent in Italian, German, Greek, etc. and were sent there to be trained as interpreters, but they ended up working in their profession.

The next morning, I went to the briefing room designated in the instructions I had received. The officer in charge greeted us and described the various schools at the Fort — General Intelligence, Order of Battle, Interrogation, Photo Interpretation, and Counterintelligence.

After he finished, he asked, "Are any of you colorblind?"

I sure didn't raise my hand.

"Don't worry, we aren't going to take away your commissions. We need colorblind photo-interpreters to look at color film aerial photos to see through German camouflage. It's urgent; we'll give you a two-month crash course and send you to England."

What the hell, I thought, and raised my hand. After two months, I became a qualified photo-interpreter; however, I couldn't see through camouflage. One had to be totally colorblind, and I was only red-green.

Me in my uniform, 1945

As I waited to be sent overseas, I received a call from Fort headquarters and reported to a room with the words Counter Intelligence Corps (CIC) printed on the door. The person at the desk was dressed in an officer's uniform and introduced himself as Major "so and so." I looked at his uniform, *no indication of rank.* He picked up my quizzical look immediately. He said that in the Counter Intelligence Corps there is no indication of rank on the uniform, except for the commanding general of the Corps. He also told me that the special agents were referred to as Mr. and wore an Army civilian insignia: **U.S.**

The major went on to say that he had reviewed my records and noted that I had studied German in college for four years. He had asked to interview me as a possible candidate to attend the CIC School for a class lasting 16 weeks. He said that the powers-that-be were now certain that we would win the war, and more CIC special agents were needed to hunt down and capture German war criminals. The other duties would be to detect, identify and neutralize the German intelligence and security services. He also told me that all this was voluntary.

It sounded more exciting than being a photo-interpreter. I volunteered and told my parents of my decision to become a CIC special agent and all that it entailed. I attended school at Fort Holabird on the outskirts of Baltimore. I passed with flying colors and returned to Fort Ritchie to await orders to go overseas.

I then wrote a letter to my parents with a return address: Mr. Bernard Marcus, APO, etc. Previous letters had a return address: Lt. Bernard Marcus, APO, etc. When I called my parents the following weekend, my mother told me that the mailman brought my letter up to their apartment and said, "Your son has lost his commission. Has he been kicked out of the Army?" She did not tell him about the CIC and told him that the Mr. was probably a mistake. She later told me that the mailman always looked at her with a suspicious eye.

The Measles

Background: This is February 2015, and there appears to be the making of a measles epidemic. I've been asked by many people, "Did you have the measles as a kid?" "No," I would say, "I had it in my twenties." The response is generally, "Oh my."

It was January 1946. I was travelling with a convoy from Le Havre, where I landed, to Paris and on to Frankfurt to Counter Intelligence Headquarters. When we arrived in Paris, the CIC officers were billeted in a very fancy French hotel. I was thrilled to be in Paris in a fabulous suite. I was getting ready for a tour of the great city when one of the fellows said, "Bernie, take a look in the mirror." I did, and saw that my face was blotchy.

I called the front desk and found out where the nearest medics were. I was examined, diagnosed with the measles and sent to a hospital outside of Paris where I was quarantined with a group of measles-infected fellow soldiers. I was sick with the measles, followed by pneumonia. This kept me in the hospital for about three weeks. I was then sent on to Frankfurt and finally Garmisch. So much for the sights of Paris and the *Folies Bergere*.

The most memorable event in the hospital occurred when a Red Cross nurse came by as I was recovering and offered me a cigarette. "I don't smoke," I said. "Try it, it's good for you," she said, "and it will help you relax." I did. It did.

When I left Germany in September 1946, I was smoking three packs a day.

Counter Intelligence Corps—"A Tough Tour Of Duty"

When I arrived in Garmisch, I reported to the CIC building and assumed my position as Agent-in-Charge of the Southern Bavarian Counter Intelligence District. Our offices were in the former Gestapo headquarters building, and our detachment numbered twelve agents. It was the rule throughout Germany that each CIC detachment was housed in its own separate quarters. We commandeered a *Gasthaus* (inn) in Grainau, which became our sleeping and eating quarters. Grainau is a small town a few miles from Garmisch on the way to the Zugspitze, Bavaria's most famous ski resort.

Gasthaus Hirth, our home away from home

Garmisch, having been the premier R&R (Rest & Rehabilitation) center for the German SS officer corps, naturally became the R&R center for the U.S. Army in Europe. The major difference was that it was available to all ranks.

The city and its ski resort were opened in January 1946. The photo shows me on my way to my first skiing experience, which turned out to be a big flop. It was about 15 years later that I learned how to ski at Squaw Valley.

My first ski trip

The CIC officer in charge of all of Southern Germany was Alan Dinehart, Jr., the son of a famous Broadway and Hollywood actor of the 1930s.

Me with Alan Dinehart, Jr. in our lederhosen

Our main job was finding, arresting and interrogating German war criminals. We had a list of those we knew had committed war crimes. They were principally SS officers and Nazi government officials who had been involved with concentration camps. The Bavarian Alps contained great hiding places for them. The German citizens who turned in the "bad guys" were rewarded with lots of food. Those caught were tried in U.S. military courts. Many prominent Nazis were arrested, with one exception, Martin Bormann, Hitler's No. 2 man at the end of the War. We would chase all kinds of leads to no avail. Rumors were that he had escaped to South America. In late 1946, he was tried in Nuremberg *in absentia* and sentenced to death. He was never found.

Another famous person we were after was Eva Braun. We were unaware that she had committed suicide with Hitler in Berlin and were chasing all kinds of leads. Late one night I received a call from a so-called informer that Eva Braun was at her sister's house about 25 miles east of Garmisch. I rounded up a squad of MPs and drove to the house. I had the MPs surround the house, and I started to climb in through a window that I was able to open. As I was getting into the room, a large German shepherd grabbed my leg and wouldn't let go. I had no alternative but to shoot the dog with my pistol. The house was empty. I wasn't injured as a result of enemy action — no Purple Heart.

In my travels around Garmisch and to attend meetings in Munich and Frankfurt, I drove a fancy BMW convertible that we confiscated from the president of BMW, who had a resort home in Grainau. We were told that Hitler used it whenever he visited Garmisch. I'm sure he "turned over in his grave" whenever I stepped into that car.

Me driving the BMW Convertible also used by Hitler

In February 1946, I was called to a CIC staff meeting in Frankfurt at US Army Headquarters. We were told that: "Our former enemies are now our friends, and our former friends are now our enemies." *Was this the beginning of the Cold War?*

In other words, we were to stop chasing German war criminals and concentrate on Soviet spies in our areas of responsibility. Intelligence sources had discovered that the Russians were infiltrating various DP (Displaced Persons) Camps, particularly those with Poles and Ukrainians who refused to be repatriated for fear of being imprisoned or executed by the Communists. It was intended that these DPs would eventually be given refuge in one of the British Commonwealth countries.

There was a DP camp near Augsburg that housed a Polish Army group, Ukrainians, and other Balkan people. It was formerly a German Army base. I was told to contact the MP officer in charge of the camp and the Polish general to inform them of our new policy and ask for their assistance in locating any suspicious persons. I was also told to spend as much time as necessary in the camp in civilian clothes and to live in the Polish officer quarters. We were to identify suspected Russian spies before Valentine's Day (not much time). On Valentine's Day, MPs would swoop into the camps and arrest the spies. They would then be brought to a prison in Frankfurt. I went back to Garmisch and changed into the clothes I had made for me by a local tailor, who did too good a job. I guess he was afraid not to. I looked too good for the DP camps. CIC policy was that all of its agents wear civilian clothes except when we attended meetings with military personnel in Frankfurt.

In my civilian clothes made by a local tailor

I was dropped off at the camp, and the MP officer-in-charge introduced me to the Polish general. The general told me that he knew of a Ukrainian priest who seemed very suspicious to his intelligence officers. The priest lived in a hut on the base and would greet some of the new Ukrainian arrivals and spend a great deal of time with them.

I arranged with the MP officer to allow the priest to have some time off in Augsburg. A Polish intelligence officer and I surreptitiously broke into his hut. We searched for anything suspicious.

We found an official-looking letter hidden in one of his robes, which was signed by someone in the name of Stalin. The Polish officer translated it. It ordered this priest to find his way to Canada and report to the Russian Embassy in Ottawa and show them this letter. The priest was to be supported in his undercover work. That was enough for me. We replaced the letter and left. On the 14th, the MPs arrested the priest and took him to Frankfurt.

Several days later, I received a call from CIC headquarters in Frankfurt to report to the CIC General's office the following morning. When I arrived, I was told that we had a problem with the British. I was to come with the General to the British Intelligence General's office for a meeting. When we arrived and were shown in, there, seated next to the General, was the Ukrainian priest. It seemed that this man was a British agent who had been living in the Ukraine for many years. We were told that at the end of the War, the Russians wanted him to relocate to Canada and spy for them. I screwed up everybody's plans. The British knew he was in this camp, and our "Priest" was meeting with other Russian agents who were told to report to him. He was then going to identify them to the British. At the end of the meeting, the British agreed to let us know of any other of their agents in the American Zone of Occupation.

I had other dealings with the British — none of them pleasant. They always treated us like amateurs, which, by the way, we were.

The Nuremberg Judges

It was sometime in early February 1946 when I was informed by the Chief of Staff of the 10th Armored Division, the local occupying force, that General Eisenhower ("Ike") was going to bring the Nuremberg Trial Judges and the Prosecutors to Garmisch for a long weekend over George Washington's birthday. Further, he said, Ike wanted to have a dinner-dance on Saturday night at the Senior Officer's Club, and he wanted a ballroom dance band and about 15 to 20 young women for the event. The women had to be over 18 years of age, speak English, be attractive, well dressed, and have no VD. Further, the band and the women were not to have been Nazis. It was my job to find a band and the women and make arrangements with the MPs to escort them to and from the Club. I was also to pay them well out of the CIC slush fund.

Where do I start? I decided to call in the local German Chief of Police and present him with this challenge. He requested some time to see what he could arrange.

Several days later he came to my office and said: "You can have the band and 20 women. Pick them up at the jail on February 23. All the requirements you had can be met with one exception. Considering that Garmisch was the R&R city for the SS, the band and women were probably all former Nazis or Nazi sympathizers."

I replied: "OK, forget the Nazi part, make the arrangements and make sure you impress on the people, no politics. You have two weeks and the people will be well paid."

On February 23, 1946. I met the MPs at the jail. We loaded everyone into four army trucks. I rode with the MP Captain in the lead jeep as we wound our way through the streets of the city with our "precious cargo" and dropped them off at the officer's club. The MP Captain and I stayed as onlookers. I was introduced as the officer who made all the arrangements to Senator Jackson, the

Chief Prosecutor for the Allied Forces at Nuremberg. He thanked me and said that this was a very pleasant break for the judges and the prosecuting staff. The party ended before midnight, and we returned our charges to the jail where we paid them and thanked them.

Time marches on. It was February 22, 1947. I was married to Molly, and we were living in an apartment in my in-laws' house in Brooklyn. My father-in-law called me to come upstairs to his apartment and showed me *The New York Daily News*. There was an article that reported that the Counter Intelligence Corps had supplied "Ladies of the Night" to the Nuremberg Judges the year before in Garmisch-Partinkirchen. He told me that he was going to burn the paper.

After all these years I forgot about this episode until February 1, 2008, when my memoirs teacher, Sylvia Halloran, pointed at three men in her class, including me, and said "There sit the Nuremberg Judges."

Zeiss

Background: After 59 years Zeiss suddenly reappeared in my life. In 2011 my wife, Ruthie, and I moved into the Vi, a senior community. We enjoy the community very much. Residents share their lives, often discussing their past and families at the dinner table. One of our fellow residents is Marianne Marx, a former professor at Stanford University. She spoke about her son Dan, who is an executive with the Zeiss division in Dublin, California. How amazing! This brought back my memories of the Carl Zeiss company in Germany after World War II ended, and my association with Zeiss scientists in the United States.

After the War, there was a rearrangement of areas between the Russian, American, British and French zones of occupation. The city of Jena was to be turned over to the Russians. Most importantly, the Carl Zeiss complex was in Jena.

I was one of those assigned to determine whether Zeiss scientists and engineers who wished to be evacuated from Jena to the American Zone were members of the Nazi party. Those who were Nazis would not be evacuated. The MPs were also instructed to take as many instruments, lenses and cameras as they could load onto trucks from the factory in Jena into the American Zone of Occupation.

I was relieved of active duty in October 1946. In February 1947, I applied for and was given a job as a research physicist to work for the U.S. Army Signal Corps Laboratories in Fort Monmouth, New Jersey. I was assigned to the Photographic and Optics Branch and shared an office with two German scientists, both from Zeiss. One was Dr. Karl Leistner. I cannot remember the name of the other scientist; we'll call him Dr. X. Further, I was told that the Laboratories were in possession of several thousand lenses acquired from Zeiss in Jena. These two gentlemen and I were to determine which would be suitable for military use.

One of these was a 210-degree wide-angle lens, originally intended for use in meteorological photography. The following two photographs are ones I took of the Dome of the Capitol in Washington, D.C. These photographs were published in the September 1948 issue of *U.S. Camera*. The first uses the 210-degree wide-angle lens while the second one uses a standard lens to show the difference.

Photograph of the Dome of the Capitol with the 210-degree wide-angle lens

Photograph of the Dome of the Capitol taken from the identical spot with a standard 48-degree lens

Dr. Leistner and I designed and had a lens-testing bench built in 1948 to help us analyze the Zeiss lenses. The article describing the bench was published in the *Journal of the Optical Society of America*.

One day in 1948, the chief of our branch informed Dr. Leistner, Dr. X and me that Professor Emanuel Goldberg, the former president of Zeiss Ikon, was coming to visit the laboratories the following week. Our chief said that he would bring Goldberg to our office for a review of our work with the Zeiss lenses.

Dr. Leistner was very excited. Dr. X was silent. Dr. Leistner told me that he had not seen Professor Goldberg since the Nazis forced him out of Germany in 1933 because he was a Jew. He said that he was looking forward to seeing him again. The day Professor Goldberg arrived, Dr. X called in sick. I will never forget when Goldberg and Leistner met. They hugged and cried. We left them alone for several hours. Dr. X left the United States soon thereafter and returned to Zeiss in Jena, then part of East Germany, for an executive position.

In November 1951, at the height of the Korean War, several of us in the Photo Branch met with the commanding general of Fort Monmouth. He told us that the General Staff in Washington was concerned that the Russians might cross the borders into West Germany as a diversion to the Americans. This would then help the North Koreans take over South Korea. He said that he was instructed to provide aerial and ground photographic surveillance of the German-Czech border. I was made the program manager to assemble cameras and organize and train a Signal Corps platoon to leave for Germany in early 1952.

One of the Zeiss lenses we had was a 100-inch focal length. We needed a camera to go with it. I was sent to Eastman Kodak to discuss this with their engineering staff and to ask them if they could build us a camera with that lens in time for the deployment of our platoon. One of their engineers was formerly at Zeiss. He said that they could, and he would like to be in charge of the design and development. Kodak management agreed. There was no contract. If they completed the project in time, they would charge the government their cost.

The camera was completed and was ready to be deployed with the platoon in February 1952. Later that month, the Fort Monmouth general, General Lawton, called me into his office to inform me that I was being called to active duty to be in charge of the platoon. We flew into Grafenwehr, Germany, and began our surveillance of the Russians. I returned to the Laboratories six months later, where I stayed until 1955. Dr. Leistner remained in the Photo Branch until he retired, and we kept in touch until he died.

Recently, Mrs. Marx introduced my wife and me to her son Dan. At our meeting, I found out that Zeiss in Dublin, California, was known for its medical instruments, including ophthalmic equipment designed by William Humphrey. What a small world!

My company, Mark Systems, designed and manufactured a wide variety of photographic and optical equipment, mainly for the armed forces. One of the devices was a stabilized binocular, which had been patented by Dr. Luis Alvarez and William Humphrey.

Zeiss has been in the forefront of optical design since the company's founding. The work of its scientists and engineers is responsible for outstanding products now produced internationally. In retrospect, we should have offered the stabilized binocular technology to Zeiss to come up with a consumer product.

Never Again

Background: January 27, 2015, was the 70th anniversary of the Soviet Army's liberation of the Auschwitz Concentration Camp from the Nazis. It brought to mind my short involvement with the Dachau Concentration Camp. The city of Dachau is located about ten miles northwest of Munich. On April 30, 1945, the US Army liberated Dachau prisoners who were then relocated to German barracks or hospitals.

I was stationed in Garmisch after the War and was responsible for counterintelligence operations from south of Munich to the Austrian border. Initially, our orders were to locate and arrest Nazi war criminals and the Nazi SS. The SS were responsible for guarding and exterminating the prisoners in the concentration camps.

In early 1946, a few CIC officers and I were ordered to assist other CIC officers in Dachau in locating and arresting the SS who had been assigned to the Dachau camp. I drove to Dachau, about two hours from Garmisch.

When I arrived in the city, there was a strong and strange odor. It became more intense as we were shown around the camp. It was remarkable — hardly any of the residents noticed the odor (we felt that they were accustomed to it). Did they have any idea of what had been going on in the camp? When asked, most of them said it was a training camp for SS troops.

We had the Military Police set up random roadblocks throughout the city, especially around food markets, so that we could interrogate residents. Since food was still very limited, we provided food for anyone who led us to the location and arrest of an SS trooper. That worked well. After two weeks in the city of Dachau, I returned to Garmisch.

The next time I saw Dachau I was standing in the Holocaust Museum in Washington D.C., viewing photographs.

My son Victor visited Dachau in 2005 and stood in the crematorium and vowed as had I 60 years before, "Never Again."

I'm Finally Home

Just Molly and Me

Sometime in the spring of 1945 my sister Millie told me of a very attractive coworker of hers to whom she would like to introduce me. Her coworker, Molly Cohen, had already agreed to meet me for lunch. I had a day and a half off every seven days from Fort Ritchie and would come home by train from Baltimore every other week. Millie and Molly worked at *Lord and Taylor* on Fifth Avenue. I agreed to meet Molly on my next trip home. I was to meet the girls at the entrance to *Lord and Taylor* at lunchtime. Millie would introduce us and take off. Millie told me that Molly was a Sephardic Jew and her family spoke Ladino, the Spanish of the fifteenth century rather than Yiddish. They were descended from those Jews who had been driven out of Spain into the Turkish empire in the late fifteenth century.

We met. Molly was very attractive, we "hit it off," and agreed to meet again on my next time off. I was to pick her up at her house. She gave me subway directions to her house. Starting in the Bronx, I had to change subway trains twice and it would take me about an hour. As planned, I arrived at her house, which was a two-story private house in the East New York neighborhood of Brooklyn. Molly met me at the door, and we took off for Manhattan for dinner by subway train. This went on for about a month.

We talked a lot about growing up and found many similarities. In fact, politically we were on the same page. Molly then said that she would like to introduce me to her parents, who wanted to meet me, but to remember that her parents and grandmother, who lived with them, spoke Ladino and very little English.

71

We met. The essence of our conversation was when was I going to go overseas and what was I going to do for a living when I got out of the Army. I responded that I had no idea.

It was evident that Molly's parents knew that we were in love and were concerned about her involvement with an Army man, albeit that most of her male friends and young relatives were in some branch of service.

In October 1945, in a restaurant in Manhattan, I proposed to Molly, gave her a ring, and asked her to marry me. She accepted, and we agreed to get married after I left the service.

My parents were thrilled. I think Molly's parents were disappointed that she hadn't chosen a Sephardic young man.

Soon after, my mother had an engagement party on an early Sunday afternoon at our apartment in the Bronx. My grandmother was there. Molly's immediate family was invited. This included her mother, father, brother Ely, sister Rachel and her grandmother.

Following introductions, I could see a strange look on my grandmother's face when she heard Molly's parents speaking Ladino to her grandmother. Following lunch, Molly's mother and grandmother lit up cigarettes with her father. I could see the shock on my parents' and grandmother's faces. A woman smoking at that time was a no-no. My grandmother then waved at me to follow her into another room where she had made me swear on my grandfather's grave that these people were Jewish.

I left for overseas in early January 1946 and came back home in September. Molly and I agreed to be married as soon as it could be arranged. The wedding would be held on October 27, 1946, my mother's birthday, in a Sephardic Temple in Brooklyn. Prior to the wedding, I was told of a very important affair to be held by Molly's mother — the showing of the trousseau of the bride-to-be to women only. My mother and a cousin were invited. I brought them to my future in-law's house and heard loud gypsy-like music while still outside. I took my mom into the house. There were about 20 women all dressed like gypsies, dancing to what I was told was Ladino music. They were relatives and friends of the Cohens. Hanging on the walls around the house was the bride's trousseau, including undergarments. I was quickly ushered out of the house and told to wait in the backyard until all the festivities and lunch were over. I was served outside after everyone else.

Lunch was traditional Sephardic food, which my mother would not eat (such as salads with Greek olives and feta cheese, dolmas and eggplant casserole).

Molly and I were married by a Sephardic rabbi, who spoke in Hebrew and Ladino, which none of our family could understand. Following the rabbi's blessing, the dancing began.

After the wedding Molly and I left for the Pocono Mountains in Pennsylvania for our honeymoon.

There were no long-time relationships between the families except when our kids were born and Debbie was married.

Our wedding picture

Fire, Fire!!!

Since apartments were hard to come by, we were living in the ground floor apartment of my in-laws' house at 704 Hinsdale St. in Brooklyn. We had a very comfortable, nicely furnished two-bedroom apartment with a living room and kitchen.

I was working in the Signal Corps Laboratories at Fort Monmouth, New Jersey. The working hours at the Labs were from 8 a.m. to 4:30 p.m. It was a tiring 2-½ hour commute each way by subway, ferry, railroad train and bus. To arrive at work in time, I woke up at 4 a.m., took the subway at 5 to Pennsylvania Station and boarded a train at 6 to Red Bank, New Jersey, arriving at 7:30. There I took a bus to Fort Monmouth, arriving just in time. Then there was the trip home. I was tired.

A buddy at the Labs had just moved into an old mansion, about 20 minutes from the Labs, that had been converted to apartments and was renting to veterans only. He told me that the attic apartment of the mansion at 752 Ocean Avenue in West Long Branch had become vacant.

I went home and told Molly about it. She loved living in her parent's house. The available mansion apartment was on the third floor. It was small compared to our Brooklyn apartment, but it was right off the beach and had a great view of the ocean. It consisted of a living room-kitchen and one bedroom. How could she leave her comfortable apartment, her family?

"Molly," I said, "I have a great job, but I can't continue this commute much longer. Please let's move, and I promise you that one day I will take you on a trip around the world." She agreed to the move across the Hudson River. We bought our first car, a 1937 Dodge, and left Brooklyn in July.

We made lifelong friends in that house—the Rostens and the Fishbeins.

On a cold, snowy New Year's Eve, December 31, 1947, there was a great party in the large dining room on the ground floor of "752." After we celebrated the New Year, Molly and I went up to our apartment and to bed. We were awakened with a start when we heard people screaming: "Fire! Fire!"

I opened the apartment door, and there was lots of smoke. Molly grabbed her coat and I put on pants, shoes and a jacket. We looked around and wondered what to take with us. I picked up my wallet and the keys to our car, and Molly took our wedding picture off the dresser. We ran into the hallway. The firemen were already downstairs and one of them shouted for us to go down the slide at the hallway window. We opened the window and there was a rubber slide all wound up, of which I had been completely unaware. I released it and the slide tumbled down three stories. It was dark.

I said to Molly, "you first."

"You're kidding," she said.

A fireman started shouting: "Hurry, come on down, I'll catch you."

Molly went first, and I followed right behind her. When I reached the ground I noticed that Molly was barefooted. She had not put on her shoes. While all of our neighbors were standing around, dressed warmly, watching the firemen hosing the ground floor, Molly was shivering. I rushed her into our car through the snow. By that time the hoses had been turned off, and the fire appeared to be out. The damage was minimal, but we were all told that the house was off limits for the time being. They told us to go to a motel or a friend's or relative's house.

I was permitted to go upstairs to get some belongings, particularly Molly's shoes. We stayed in a motel for a week. The fire inspector decided that the fire was the result of faulty electrical wiring. The landlord cleaned up the damage and paid our expenses. He also had an outdoor fire escape built. We continued to live in the apartment until December 1, 1948, a few weeks before our daughter Deborah was born on the 24th. Molly couldn't climb the three flights anymore.

In 1995, Ruthie and I drove from Washington, D.C. to New York City. On the way we drove down Ocean Avenue to look for the old house. It was gone. Nothing was in its place.

No First Name Or Middle Initial

At age 16, before I entered college, I had to show my birth certificate. It read Male Marcus. My parents could not agree on my name for quite awhile after I was born, so the doctor had written Male. My father and I went to the NYC Department of Records and explained the situation. A new birth certificate was issued. I was now officially known as Bernard "NMI" Marcus, NMI being the abbreviation for No Middle Initial. Now to the no-middle-initial story.

In the summer of 1956, one year after we moved to Pasadena, Molly and I felt that we should own our own home. The brother of Molly's friend Ruthie, Seibert Weissman, had just received his broker's license, and we hired him to help us. He found us a charming old house that we could afford in Altadena on Vistillas Rd. I went to a Bank of America branch in Pasadena to obtain a mortgage. We filled out all the paperwork and waited for several weeks. It was time for the kids to start school, and we were anxious to move in. I contacted the banker who told me that he was sorry to inform us that our loan was disapproved because we had a bad credit rating.

"What's the problem?" I asked, "I've never borrowed any money, I pay cash for everything, I have a good job. I have a savings account in your bank."

"Perhaps you should see the branch manager," he said.

He excused himself, went in to the branch manager's office and in a few minutes came out and invited me to come in.

"You have a big problem," the manager said. "You came here from New Jersey, and you're too young to be the Bernard Marcus from San Jose, California. That Bernard Marcus has been in jail for passing bad checks. Don't you have a middle name?"

"No," I said.

"Well," he said, "Pick a middle initial, go out and buy something on credit, give us as a reference, then pay it off in a month. In that way you will have established a good credit rating, and we'll process a mortgage so that you can buy the house."

Out of the air, I picked the letter P. I had my driver's license changed. I notified my company, the Social Security Administration, the VA, insurance companies, and in time, I became accustomed to the initials BPM.

Many years later, when we were living in Los Altos Hills, an article appeared in the San Jose Mercury News that Bernard Marcus had been arrested again for passing bad checks.

My Involvement In Pre-Israel and Israel, 1946 – 1967

I would like to recount some of the events that took place after the War as they relate to the Jews who were in concentration camps in the American and British-occupied zones of Germany. In the American zone, the Jews were evacuated from the concentration camps and were billeted in empty German military barracks and hospitals. They were free to come and go as they pleased. In the British zone, they were kept in the concentration camps, where they were cared for and fed, but they were not free to leave the camps.

In early March 1946, the Commanding General of the Counter Intelligence Corps (CIC) summoned me to the Headquarters of the US Army in the IG Farben Building in Frankfurt. Upon my arrival, we immediately went to the office of the Commanding General of the US Forces in Europe, General McNarney. There were two other senior officers present. The General had been informed about my rank (CIC officers did not wear an insignia of rank on their uniforms) and also that I was Jewish. He then told us that we were going to a meeting in the office of the British Commanding General in the same building, and that the British were going to lodge a complaint against the United States Forces. He said that I was to listen and not say a word.

We then left for the British General's offices and arrived promptly at 10 a.m. When we arrived, General McNarney's aide introduced us, saying that my responsibilities included the control of the border guards near Mittenwald. Following introductions, tea, coffee, and snacks, the British General read from a document, which he said was a letter from one ally to another. He requested that the U.S. Forces immediately restrain all Jews from leaving the compounds in which they were housed. Moreover, he asked that the border at Mittenwald be closed to all Jews leaving Germany.

Background: For months, tens of thousands of Jews had been leaving Germany on trains that they boarded in Munich and traveled south through Mittenwald into Austria and Italy. The American border guards had been instructed to let them go. In Italy, the refugees boarded ships that took them through a British Naval blockade to Palestine.

General McNarney took the letter from the British General and said that he would send it on to Washington for instructions. He told the British General that he would not restrain the Jews and that they would be free to come and go within the American Zone. However, he would order that the border be closed until he received instructions from Washington.

He turned to me and said "Mister, you have your orders." We left and returned to General McNarney's office.

He stood looking out of his window, and said to me "Lieutenant, Mittenwald is at least 300 kilometers from here, and I'll be damned if I can see that far." I'll never forget him or those words. The trains loaded with Jews kept rolling through Mittenwald. We never received orders to the contrary. I left Germany in September 1946.

Long after I got out of the Army and with the help of the CIA, I started my own company in 1962 — Mark Systems. Our products were all classified optical and photographic systems with one exception: an image-stabilized binocular invented by Luis Alvarez, a Nobel Laureate, and by William Humphrey.

In January 1967, a CIA officer told me that an Israeli Army officer, Brig. General Shapiro, would visit me at Mark Systems. If I agreed to his proposition, I was to call the CIA officer and I would receive further instructions. General Shapiro showed up the next day.

His story was as follows: the Israeli Air Force was in need of aerial cameras and film. The United States did not wish to appear to be partial to the Israelis and would not provide or sell aerial cameras and film to them. He said that Arab armies surrounded all of Israel and that the Israelis needed to know the strength and composition of these armies. If I agreed, the CIA would inform me of how Mark Systems could help to provide cameras and film to Israel and that the Israel Purchasing Agency in New York would reimburse Mark Systems. I agreed.

The following week, I scheduled a business trip to Washington, D.C., where I met with the CIA officers involved in this transaction. I was told that the United States would like the Israelis to have their own aerial photographic capability as soon as possible and the quickest and most covert way was for Mark Systems to act as a conduit. I was told that the following week there would be a sale of fifty "obsolete" aerial cameras at Wright Patterson Air Force Base in Dayton, Ohio, that I would be the only bidder present, and that $3000 per camera would be a fair bid. In addition the Israelis needed aerial film from Kodak. Kodak would not sell directly to the Israelis because an Arab boycott would close Kodak facilities all over the Arab world.

After purchasing the cameras, I was to go to Kodak headquarters in Rochester, N.Y. There I would meet with an old colleague who was in charge of Kodak's covert business. Kodak was already informed of the purpose of my visit. I was to give the company an order for 1000 rolls of 9-inch wide by 100-foot long film. I did all that and gave them a shipping address to a warehouse in Brooklyn. I then went to New York City to the Israel Purchasing Agency, where I collected a check for all I had spent.

The rest is history. The Israeli Air Force located the positions of the Egyptians, Syrians, and Jordanians, destroyed the Egyptian Air Force, and was successful in the Six-Day War in June 1967. As a remembrance, I received an 18"x18" print from an aerial photograph of the Eastern Wall and the Blue Mosque in Jerusalem taken the day after the war ended.

Jerusalem/The Eastern Wall and the Blue Mosque – June 1967

In addition, my family and I were invited by the Israeli Defense Forces to come to Israel.

In 1973, I was invited to participate in a meeting in Israel by Yitzchak Rabin and Golda Meir. Unfortunately I could not attend because Molly was ill; however, I participated by telephone with many others. The end result was that the Israelis made a deal with Mark Systems to manufacture several of our telephone products.

EMBASSY OF ISRAEL
WASHINGTON, D.C.

שגרירות ישראל
ושינגטון

October 5, 1972

Mr. Bernard Marcus, Chairman of the Board
Mark Systems, Inc.
10950 N. Tantau Ave.
Cupertino, Calif. 95014

Dear Mr. Marcus:

    I have the pleasure of informing you that the Prime Minister of Israel
is convening in Jerusalem the Third Economic Conference which will consider
the major economic issues of Israel today and the programs needed to develop
the economy in the next decade. The Conference will take place from May 27
until May 31, 1973. Invitations are being extended to a select group of
people from all over the world.

    The Economic Conference will be the major event of the 25th Anniversary
Celebration of the State of Israel. The Prime Minister and her colleagues
will use the opportunity to sum up and draw lessons from five years of follow-
up activities since the first Conference and set new goals for the future.

    A forum of elite industrialists, financiers and trade entrepreneurs will
gather within the framework of international professional committees to discuss
business ventures and investment ideas that will be mutually advantageous for
you and the State of Israel. The 25th Anniversary Exhibition encompassing
Israel's achievements in trade and industry will be inaugurated and open especially for the attending delegates during the Conference.

    The Prime Minister has asked me to convey this advance information to you
in the hope that you will be able to attend the Conference. After receiving
your confirmation, the Prime Minister will send you a personal invitation.

    We look forward to your positive response.

Sincerely yours,

Y. Rabin, Lt. Gen. (Res.)
Ambassador

R.S.V.P.
Mr. Pinhas Rimon
Government of Israel, Investment Authority
850 Third Avenue, Suite 604
New York, New York 10022

Letter from the Embassy of Israel

85

Jerusalem, March 8, 1973

Dear Mr. Marcus:

I am very pleased to learn that you have accepted our cordial invitation to participate actively in the Third Economic Conference, a central event in Israel's Twenty-fifth Anniversary Year. This historic milestone, marking a quarter of a century of Israel's reestablishment, is a remarkable tribute to the power of a people's faith and dedication, as expressed over the preceding nineteen centuries. Though Israel's path to peace and prosperity has not been an easy one, we have great cause to celebrate the unique accomplishments of our restored statehood, and to plan together for our economic future.

Next to peace itself, economic development and social progress are major Israeli goals. Your active role in the transfer of capital, technology and operational know-how furthers the universal aim of all men who wish to see a prospering Israel in a world of cooperation and flourishing trade. We have learned and can yet learn much from our good friends from overseas. Your personal involvement can help make Israel an economic and social model for others.

I look forward to greeting you personally, on May 27, in Jerusalem, the symbol of the highest aspirations of the Jewish people and of mankind. Out of our meeting, new dimensions in our mutual cooperation, for the good of all involved and for Israel, should flourish and grow.

Sincerely yours,

Golda Meir

Golda Meir

Letter from Golda Meir

Eyes In The Sky

My Association With Dr. Land

From 1950 to 1985, I met and had business dealings with Dr. Edwin Land, President of Polaroid Corporation. Some of our meetings were under unusual circumstances. I first met him in New York City in May 1950 at a meeting of the Photographic Society of America (PSA). I was chairman of the speakers committee and had received a call from a Polaroid patent attorney that Dr. Land would like to speak to our group to introduce a revolutionary new product. In addition, he asked us to invite members of the press. At this meeting he spoke about and demonstrated his new one-step photographic process, as described in the clipping in this photo.

Dr. Edwin H. Land, president and research director of Polaroid Corp., Cambridge, Mass., discusses details of Polaroid's new Type 41 film for one-step photography with members and guests of New York Section Technical Division, PSA, May 2, 1950 at Wilkie Memorial Building, New York City. New film makes black-and-white print image instead of sepia color characteristic of first Polaroid (Type 40) material. This informal discussion followed Dr. Land's formal paper on "New Films for One-Step Photography." (*Photo by John Wolbarst*)

Me at the far left, Dr, Land in the center

We next met in late 1951 while I was working at the Signal Corps Labs. The Army wanted an aerial camera utilizing the new Polaroid film. I made an appointment to meet with Dr. Land to discuss a development contract with the Polaroid Corporation. We met in his laboratory in Cambridge, Massachusetts. The building was an unmarked warehouse with a side door. The guard searched my briefcase to insure I did not have a recording device. A patent attorney was present who requested that I sign an agreement that any discussions with Dr. Land that might result in a patent would be the sole property of the Polaroid Corporation. Further, that I would not disclose any technical information I might learn that Polaroid had not yet made public. Dr. Land agreed to have Polaroid build ten aerial cameras, which I later took with me when I went to Germany in 1952.

Dr. Land's laboratory was very unusual. Other than the guard and the patent attorney, who was present at all times, the employees of the lab, including the scientists, were women. He later told me he could trust women with secrets, but not men. One of the lab rooms had a sign, which read, "Color Lab." After some prodding, he told me the rumors that Polaroid would introduce a color film were true. I told him I was red/green color-blind. He thought that was great and wanted to use me as a test subject to look at various color photos from time to time. I was to tell him which were most pleasing. His reason for having me do this was that roughly 20 percent of men were red/green color-blind, and he wanted his film to be acceptable to as many people as possible. I looked at many photos over a period of several months, until Polaroid introduced its color film in 1953.

It wasn't until September 1955, while working at the Hycon Corporation, that I met Dr. Land again. I was the project manager in charge of the development of the cameras for the U-2 airplane. A program meeting was called by the CIA to be held in Cambridge, Massachusetts. The meeting took place in the Polaroid laboratory building. Present at the meeting were representatives of the various manufacturers (i.e., Hycon, Lockheed, Perkin-Elmer). Dr. James Killian of Harvard, who designed the optics, was also present. Richard Bissell, the Deputy Director of the CIA, was in charge. He introduced Dr. Land as the "Father" of the program. We were informed Dr. Land was chairman of a special intelligence committee reporting to President Eisenhower, as well as the President's Science Advisor.

I met Dr. Land again in 1959. I had started working for Itek Corporation in Boston in July 1958 as the project manager for the camera system for the Corona satellite, a covert CIA program. In October 1959 we were having great difficulty keeping the camera together on a vibration table. The camera system was scheduled to ship by December 1 to the Lockheed Skunk Works, where the camera would be installed in the satellite. The satellite was then transported during the night to Vandenberg Air Force base, where it would be launched into space.

We reported our difficulties to the CIA project officer, who said he would report it to his supervisors. Several nights later, after some modifications to the camera, we were conducting vibration tests again. The security guard came up to me and said there was a visitor in the lobby who claimed to have the proper clearances and wanted to witness our tests. I went into the lobby and shook hands with Dr. Land. The camera withstood all the tests. We met our December 1 deadline. The Corona satellite successfully flew on August 18, 1960.

I again saw and shook hands with Dr. Land on August 18, 1985, the 25th anniversary of that launch and recovery. Dr. Land, other principal contributors from the CIA, Air Force, the contractors, and I received a special Presidential Space Pioneer medal.

The end of an era occurred on February 8, 2008 when the Polaroid Corporation announced that by the end of the year Polaroid instant film would be discontinued.

Fort Monmouth, New Jersey, 1946-1952

October 4, 2007, was the 50th anniversary of the successful flight of the Soviet satellite, Sputnik. It was also the 51st year after the first successful overflight of the USSR by a U.S. U-2 reconnaissance airplane (July 14, 1956), and the 47th year after the first successful U.S. Corona satellite flight (August 19, 1960). All three were monumental events. I was proud to be involved in the two U.S. programs.

It all started when I returned from my World War II service at the end of September 1946. My orders appeared strange. I was to report to a basement room in the Old State Department building next to the White House. Imagine my surprise when I stepped into that room and saw my former commanding officer of the Counter Intelligence Corps in Germany.

"Bernie," he said, "We're planning to start a new intelligence agency and we would like you to be part of it. If you agree, we will send you back to Germany as a civilian with good pay. Just keep on doing what you have been doing as a CIC officer."

I told him, "No thanks, I plan on getting back to my civilian job at the Fort Monmouth Signal Corps Laboratories, getting married and going back to school to get my advanced degrees. The CIC was a great experience, but I have had enough of cloak-and-dagger adventures." I told him that I would remain in the active reserves. He wished me luck and told me they would keep in touch, which they did.

After Molly and I were married, I returned to work at the Signal Corps Laboratories.

I enrolled at the Polytechnic Institute of Brooklyn, where I could get my master's degree in physics by taking night and Saturday classes. The trip took two hours by railroad train to Jersey City, ferry to Manhattan, and subway to Brooklyn. We began raising a family—Deborah was born on December 24, 1948, and Victor was born on July 31, 1950. Before Victor was born, we were fortunate to get a small-attached house in a veterans housing complex in Eatontown and learned to shovel coal into a furnace.

Meanwhile, with the outbreak of the Korean War in 1950, the Signal Corps was designated by the Army to develop an aerial photographic capability since the Air Force was no longer the Army Air Corps. The branch I was part of at the Laboratories was given that job, and I was made the project manager. The airplane to be used was the single-engine Cessna L-19. All camera equipment was to be handheld. In deference to the Air Force, no cameras could be attached to the airplane.

We had Eastman Kodak and Polaroid do the design and fabrication of the cameras, which took more than a year. At Kodak, a former Zeiss employee was the chief designer.

In March 1952, I was called to the office of Major General Kirke Lawton, the Commanding General of Fort Monmouth. He told me that I was going to be called to active duty and assigned to the Signal Corps. My orders were to go to Washington to be briefed at CIA headquarters, and then return to Fort Monmouth for further orders.

I thought that I had escaped the Korean War. Lots of things had to be put on hold. Important for my future was my schooling. I was going to Brooklyn Poly at night, working on an experimental thesis towards my degree. Molly and I were very distraught, to say the least.

When in Washington, I found out that the CIA had funded the Army for the R&D and equipment purchases for its aerial photographic capability. Also, they were responsible for calling me to active duty. I was told that the CIA believed that the Russians were considering a sneak attack across the Czech/West German border during the summer in order to help relieve the pressure on North Korea. I was to select a group of qualified Army photographic personnel and leave for Germany with our new equipment in early June.

We were stationed in Grafenwehr on the border and were assigned two L-19s and given our own photo lab. I reported to a CIA officer, who had brought along two photo-interpreters. We were responsible for covering about 50 miles of the border and took aerial and long-range ground photos continuously for two months. The border was clearly marked. The Russians had erected two barbed wire fences separated by a clearing of about 100 yards all along the border. We didn't dare cross over in our little airplane because the Russians were watching us closely.

At the end of August, with no movement across the border, we were sent home. I reported to Signal Corps headquarters in Washington, where I was told that as a result of our efforts the Army would have its own reconnaissance capability run by the Signal Corps. In addition, the Army was to have a reconnaissance drone, and I was to be in charge of that R&D effort at Fort Monmouth, not as an officer, but as a civilian. That was the end of my Korean War duty, and it was a great relief to get back to my family and school.

The Communist Menace, 1952-1954

It was August 1952 and I was back in Fort Monmouth after my duty in Germany. Molly was delighted to have me home. Caring for our two small children alone—Deborah, 3 and Victor, 2—exhausted her. If we had lived anywhere in New York, including the end of Long Island, she would have had help from the family. Since we lived in New Jersey and the Hudson River seemed wider to cross than the Pacific Ocean to New Yorkers, she took care of everything herself.

Back at school, I was devastated to learn my thesis experiment had been dismantled because no one expected me back before the Korean war ended. A year's work was lost. I began to reconstruct my experiment.

Meanwhile, back at work, I wrote the specifications for a drone reconnaissance system and selected the contractors for the drone and camera.

Everything was going smoothly until a Friday afternoon in mid-October 1953, when I received an unexpected visit in my office from General Kirke Lawton, the commanding general of Fort Monmouth. He closed the door and said that what he was about to tell me was Top Secret.

- The convicted and executed Communist spy, Julius Rosenberg, a City College of New York (CCNY) graduate, had set up a Communist spy ring at the Laboratories where he had worked as an engineer from 1940 to 1945.

- Fort Monmouth was a "House of Spies," and he had invited Senator Joseph McCarthy to come to Fort Monmouth to conduct hearings that would alert the country to this "Communist Menace."

- The FBI had identified 34 engineers who were "Communist spies," and there were 80 more who were still under investigation. They were all graduates of either CCNY, the University of Michigan, or UCLA, and all but one were Jews. He said that the 34 "spies" were already under suspension. The following Monday all guards would have the list of the 80 under suspicion, who would be escorted to a facility outside of the Laboratories. There they would be informed that their security clearances had been revoked pending hearings to be conducted by Senator McCarthy. The hearings were to begin the last week in October. A memorandum would be issued to all laboratory personnel informing them of these events.

- Finally, he said that he had the Department of the Army recall me to active duty effective immediately to be Senator McCarthy's aide while he was at Fort Monmouth. He handed me my orders. I was to ensure that the Senator and his staff were properly welcomed and had everything they needed. All senior people at the Laboratories and his staff would be informed that afternoon.

"Why me?" I asked.

"It's simple," he said. "You went to City College, you're Jewish, you are a Counter Intelligence Officer with top secret clearance, and the FBI said that you are OK. We don't want anyone to accuse us of being anti-Semitic."

I was shocked. Many of the anti-Semitic jabs I experienced when I was younger flashed through my head. I became concerned about which of my friends might be involved.

Right to left: Me, General Lawton, and others waiting for Senator McCarthy's arrival at Fort Monmouth

The next couple of weeks were hell. Senator McCarthy never spoke to me. His assistants, Roy Cohn and David Schine, gave all the orders or requests. The most bizarre was to provide a small refrigerator behind the curtains on the stage that had been set up for the hearings, or more aptly, the interrogations. A pitcher of gin martinis was brought in every day by David Schine and placed in the refrigerator. Schine's job was to keep the Senator's "water" glass full.

Two of my best friends were suspended, and another had his security clearance lifted. The relationships we had enjoyed became strained. The "hearings" went on for two weeks. Many of the suspensions had to do with the parents of the accused. For example, several parents had become members of the Workman's Circle when they arrived in the United States in order to obtain cheap burial sites. That organization was on McCarthy's list as a Communist front. By the senator's definition, the sons were as suspect as the fathers.

Those of us not affected, though friends of the "spies," were concerned about guilt by association, and so we avoided seeing each other. Our wives kept in touch by telephone. The only subject was "How are the kids?" At the Labs, hardly anyone spoke about these events, especially around me, because after all, I might report them to Senator McCarthy. It was a nightmare. At home we stopped reading the *New York Post*, a liberal newspaper at that time, and started reading the *Herald Tribune*.

Tension built up until the McCarthy hearings in the U.S. Senate began several months later in 1954. Relief set in with his censure. Soon after, General Lawton was relieved of his command. He came to my office to say goodbye and to warn me that the Communist infiltration of the Labs was real.

However, by the end of 1954, all 34 of the suspended engineers had been reinstated. The other 80 had their security clearances restored. All but one went back to work at the Labs. So much for the Communist menace. Unfortunately, I had lost many good friends forever.

Fort Monmouth to Pasadena, 1954-1955

The reconnaissance drone and its camera were completed in December 1954. The manufacturers were Radioplane and Fairchild Camera. Field trials were scheduled to be held at Camp Irwin in the California desert in February 1955. At Brooklyn Poly, I finished my thesis and started preparing for my oral exams scheduled for the end of March.

The CIA funded the drone program. Its project officer, a Signal Corps major, would always come to our home, not to my office at the Labs, for an update. His reason was that he knew some of the officers at the Labs and did not want them to know of his agency assignment. Molly called him "The Spook." When he visited in January 1955, he told me that the CIA had started a major new program. A contract to design and manufacture the cameras and provide field service support would be placed with the Hycon Manufacturing Company in Pasadena, California. He also said that an executive vice-president of the company, Bill McFadden, would call me, invite me to visit, and offer me a job. He went on to say that the program was highly classified and of vital national importance and encouraged me to seriously consider what would be offered.

Soon afterward, I did receive a telephone call from Mr. McFadden and agreed to visit Hycon after the field trials at Camp Irwin were completed. The field trials were successful and, with that behind me, I drove to Pasadena. The mountains were beautiful and sharp, and the air was cool and clear. When I met Mr. McFadden, he told me all about himself. He was a mechanical engineer from Cal Tech and had worked many years for Walt Disney.

He said that he learned a lot about me from an unnamed customer and told me that the contract Hycon would receive was highly classified and that very few people in the company were cleared. He was putting a team together and was informed by his customer that I would be a good choice for program manager. I spent several days getting to know Mr. McFadden and several of the other officers, and vice versa. By the end of the week, I was offered a job at four times my salary at the Labs plus stock options, about which I knew nothing. Also, the company would pay for our moving expenses and our trip across country and give us time to find a place to live. If I accepted, I would have to arrive for work on August 1. I told Mr. McFadden that I would discuss this with my wife and let him know within a week after I returned home.

Molly and I thought this would be a great adventure and a way to escape the gloom at Fort Monmouth that still prevailed after the McCarthy era. I accepted the offer. In March, I passed my oral exams and in May graduated with an M.S. in Physics. We left New Jersey in mid-June in a 1952 De Soto with no air conditioning for a three-week trip across country. As we arrived in San Bernardino, Molly, my sister Elaine (who came with us), Deborah, Victor and I all complained that our eyes were burning and tearing. I told them that the mountains were on the right, very high and nearby. No mountains to be seen. We were experiencing our first smog.

We settled in a motel in Arcadia. I left Molly to look for a place to live, while I reported to work in an unmarked building in Pasadena. A CIA security officer, Walt Lloyd, who has remained a close friend to this day, promptly briefed me.

"Welcome to the U-2 program," he said.

The Balloon Program

In the early 1950s, the Rand Corporation in Santa Monica and the Boston University Physical Research Laboratory, under contract with the U.S. Air Force, originated the concept for the "Balloon Program." The food company, General Mills, of all choices, was given a contract to develop a nitrogen-filled balloon with an attached gondola to carry two reconnaissance cameras. One was the Hyac camera; the second was a 16mm movie camera. At Hycon Manufacturing Company, we built 500 Hyac cameras in 1955. At the same time in another secret facility, unknown to the Hyac group, we were developing the cameras for the U-2 program. Being in charge of both meant a seven-day workweek for me and making sure the right hand never knew what the left hand was doing. The U-2 program was unknown to Air Force personnel, except for its Chief of Staff.

The balloons were designed to fly at an altitude of 35,000 to 50,000 feet and were launched in January 1956 from sites in Norway, Scotland, West Germany and Turkey. They were to fly over the Soviet Union, carried by the winds at that altitude. Each balloon had a radio transmitter with its own ID number. They were to be spotted by recovery aircraft in the Pacific, which would then send a signal to the balloon to release the gondola. The aircraft would snatch the gondola in midair. Over 400 balloons were launched in a one-month period and 40 were retrieved. The rest were either blown up by Russian Aircraft or landed in Poland, Russia and China. Chaos! I have no idea of the value of the results because I had become immersed in the U-2 program.

Pasadena, CA, 1955-1958

I started writing this part of my story on May 2, 2008. It was 48 years ago on May 2, 1960. I was sitting in Walt Lloyd's office in a CIA building on H St. in Washington, D.C. We were discussing launch plans for the reconnaissance satellite when he received a telephone call from his boss Richard Bissell, Chief of Directorate of Plans (the CIA "black or covert programs").

"Bernie," Walt said, "A U-2 was shot down over Russia and the President needs a cover story to give to the press. I've got to get the one we've already written out of the file and run it over to the White House. We assume the pilot died in the crash."

That day President Eisenhower announced that a U-2 aircraft on a weather flight along the Soviet border might have strayed into Soviet territory and crashed. Several days later Premier Khrushchev paraded the pilot, Gary Powers, on Soviet television and released the following photo to the press.

Khrushchev and the remains of the U-2 and camera system

Now back to August 1955. Soon after Walt Lloyd briefed me on the U-2 program, we visited the Lockheed Skunk Works, where I met Kelly Johnson, Lockheed's Vice President of Research, and his design staff. Skunk Works was the name given to Lockheed's advanced research and development location located at the Burbank Airport. There, a group of engineers worked on advanced and top-secret projects, uninhibited by organizational constraints. They proceeded to give me a history of how the U-2 was conceived and how it ended up as a top-secret CIA program.

In late 1954, Kelly Johnson, who had designed the famous P-38 fighter-bomber, presented a proposal to the Air Force for an extremely high-flying (70,000 feet) photo-reconnaissance aircraft intended to fly over the Soviet Union. The aircraft would fly at a speed of 500 miles per hour to a range of 3000 miles with the pilot as the lone crew member. The Air Force, however, was not interested since it had a competing aircraft being designed. The Lockheed proposal came to the attention of the President's Intelligence Committee headed by Dr. Edwin Land. With President Eisenhower's approval, he went to the CIA and spelled out a new approach for the Agency to gather intelligence by using high-flying aircraft that would supposedly be invisible and immune from being shot down.

By March 1955, Lockheed received a contract from the CIA and the detailed design and construction had begun. Soon after, the following contracts were issued: Harvard University was to write the specifications for the camera systems to be employed and to design the lenses; Perkin-Elmer Corporation was to build the lenses; Hycon Manufacturing Company was to build the camera systems and provide a field service crew to support Lockheed in the field; and Eastman Kodak Company was to provide a secret location in Rochester where they would develop the film from each flight. The pilots were all Air Force officers who were transferred from the Air Force to the CIA and were now known as Mr., with no indication of rank. All the aircraft would have

NASA emblems to indicate that they were weather research planes. We were told that Lockheed had just completed its first successful flight, and I was handed a drawing of the equipment bay and told that the schedule was to deliver a camera to Lockheed for test flights in the Nevada desert by March 1956.

After this session at the Skunk Works, a Lockheed pilot flew Walt and me to what was known as the "Ranch." In reality, this was a huge dry lake (Groom Lake) in the middle of Nevada where the final assembly of the U-2 took place and where the cameras would be integrated and test flights flown. It seemed like the hottest place on the face of the earth. I was shown the airplane and the camera bay and met some of the pilots.

The U-2

We then returned to Burbank and I went back to Pasadena, where I had my work cut out for me as program manager for the cameras and the field service crews. In the following eight months, we completed the design and construction of a 36-inch focal length f/10 panoramic camera named the "B" camera. It used 5000 feet of ultra-thin, nine-inch wide film and provided about 4000 aerial photos. It could resolve features as small as 2.5 feet from an altitude of 70,000 feet.

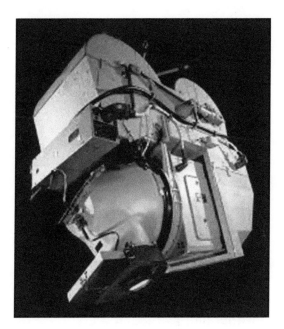

The "B" camera

We also organized a field service crew of 20 highly trained former Air Force technicians. Test flights flown over the United States in April 1956 were successful. Airplanes, camera, crew and support personnel were then sent to England. The first operational mission was in June, flying over East Germany and Poland. The British then decided that they did not want the U-2s operating out of England, which forced a move to Wiesbaden, Germany. After several flights from Wiesbaden, a German newspaper carried photos of the U-2s taking off with a story that the U.S. said they belonged to a weather reconnaissance squadron. In September, it was decided to move the U-2s to Adana, Turkey, and to Pakistan, from where they were deployed.

Note: After the Gary Powers incident in 1960, they were returned to the United States.

It was later reported that these missions substantially added to the U.S. intelligence community's knowledge of Soviet military forces and industrial capability.

During this time, I saw very little of Molly and the kids. When I was in Pasadena, I worked twelve hours a day, six days a week. My job required that I make numerous trips to the "Ranch," to Washington, and to Rochester to visit Eastman Kodak, where I looked at processed film to make sure the camera was operating properly.

Molly and I felt that the kids, as well as the two of us, needed to have a feeling of belonging, so we bought our first home in Altadena in 1956.

In December 1957, Trevor Gardner, who had been Assistant Secretary of the Air Force for R&D, returned to Hycon to become its president. During the Christmas holidays, he and I went to visit Richard Bissell in Washington to lobby for a new program (i.e., a spy satellite) about which Gardner had learned and that the CIA had been directed to build. The following excerpts from a book written by Jonathan E. Lewis titled *Spy Capitalism: Itek and The CIA* tell that story.

Page 91:

> Trevor Gardner believed that Bissell should consider Hycon. Gardner had learned that Itek was suddenly in the running to win the camera contract.
>
> Gardner felt that Hycon, which had designed the camera for the U-2 program and worked closely with the CIA to provide field support services to U-2 bases around the world, should be given a chance. He went to visit Bissell in Washington and brought along Bernie Marcus, a top executive at the company.
>
> Meeting in Bissell's office, Gardner and Marcus lobbied hard for their proposal. During the discus-

sion, Gardner excused himself to use the men's room. Bissell told Marcus that he had already made up his mind; the program would not go to Hycon. The U-2 program would continue, and Hycon would retain that business.

Bissell assured Marcus that he would let Gardner down easily. Then, before Gardner returned, Bissell gave Marcus a proposal to consider. He asked him to leave Hycon and get a job at Itek. The next time Marcus met with Bissell on business for Hycon, Walter Levison, a vice president of Itek was waiting outside the office. Within weeks, Marcus was working at Itek.

Page 112:

In meetings with Walter Levison, Bissell received a briefing on Itek's design concept and discussed the assignment of Bernard Marcus as Itek's first Corona Project Officer. Marcus, who had been recruited to Itek from Hycon, the manufacturer of the U-2 program's spy cameras, was given the assignment because he could be "easily cleared" by CIA security, and because "Mr. Bissell highly recommended him."

In July 1958, three years after Molly, the kids and I left New Jersey for Pasadena, we were driving east to Boston, home of Itek. The company was founded in December 1957, with Rockefeller money buying Boston University's Physical Research Laboratory, all arranged by Mr. Bissell. Itek's offer, we felt, was very generous. My salary was doubled, I would be paid for overtime and receive stock options. Molly was also eager to return to the East Coast to be closer to her family.

Epilog:

- *Several months after his capture Gary Powers was convicted by the Russians of being a spy. In February 1962, he was exchanged in Berlin for a Russian spy the United States had captured. He came back home and worked for Lockheed as a test pilot from 1963 until 1970. He then broadcast traffic conditions from a helicopter flying over the Los Angeles freeways. In 1977, the helicopter he was flying lost power and crashed. Gary Powers was killed instantly.*

- *Ruthie and I were invited by the CIA to attend a U-2 reunion in a classified CIA facility in Washington, D.C., on July 4, 1996, the 40th anniversary of the first flight of the U-2 over Russia. It was a gathering of the key players from Lockheed, Hycon, the CIA and the Air Force. The wives were out on tours of the city, lunch, etc. There was a lot of reminiscing, followed by surprise speakers. The first was Gary Powers' son, who spoke about his dad after the U-2 incident, and reiterated that Gary had not given the Russians any secrets. The next speaker, and a real surprise, was a Russian general. He was introduced as the officer in charge of the Russian antiaircraft division, which for four years had been given the responsibility of shooting down the U-2s as they were flying over Russia. He spoke in Russian with a translator. He said that Russian fighter planes could only fly several thousand feet below the U-2 and were incapable of shooting them down. The Russians knew when every U-2 took off. A new surface-to-air missile had exploded near Gary Power's plane and brought it down. The Russian said that he was promoted to general soon thereafter.*

- *In October 2015, Ruthie and I went to see a movie "The Bridge of Spies." The movie accurately related the shooting down of Gary Powers by the Russians and his subsequent release. There was also a quick look at the remnants of the U-2 as well as the "B" camera, which survived the crash. I must say that I felt proud that I had some part of this moment in history.*

Only in America!

Boston/Palo Alto, 1958-1962

After a two-week drive across the northern part of the country, we arrived in Boston in early August 1958 and moved into a motel in Lexington. In discussions with Richard Leghorn, President of Itek, he told me that the company, now in Boston, would be building its headquarters in Lexington, Massachusetts. This convinced Molly and me that we should look for a house in that city. Molly and the kids did the looking while I started my job. She found a great new house on Baskin Road, within walking distance of the grammar school and the town center. There were no fences between the houses and lots of trees. What a difference from Southern California! September came, Debbie and Victor started school, Molly started to furnish our new home, and I started to learn all about satellites and the origins of Itek.

Itek was started in November 1957 with funding from Lawrence Rockefeller and the involvement of Richard Bissell of the CIA. Itek acquired the Boston University Physical Research Laboratory in January 1958. It gave the company an instant research and development capability in photography and optics. The CIA then awarded Itek a contract to work with Dr. James Killian, President of MIT, and Dr. Edwin Land on the feasibility of a panoramic camera to be flown in a stabilized satellite. A key feature would be a film transport system that would transport film to a capsule that would detach from the satellite and be recovered by an aircraft.

In January 1958, a meeting was held in a CIA office in Washington, D.C., to kick off this country's first satellite program. It was highly classified and was to be a photo reconnaissance spy satellite. Richard Bissell, the CIA director in charge, told those of us at the meeting that the program needed an innocuous name and asked for recommendations. Seated at his desk in one corner of the room was the program's contracting officer, George Kucera.

He raised his hand and pointed to his typewriter. All heads turned and nodded. Mr. Bissell said: "That's it."

The Corona, the first U.S. and a highly successful spy satellite, was named after a highly successful typewriter, the Smith-Corona.

Lockheed Missile and Space Division in Sunnyvale was assigned the satellite contract and served as prime contractor, General Electric Space Division designed the reentry vehicle, and Eastman Kodak had the contract to develop the film. As with the U-2 program, time was of the essence. The program timetable was to start launching and recovering an empty capsule starting in the spring of 1959 and with a camera soon thereafter. Our contract was to complete the design, fabricate a model, environmentally test it and begin delivery of ten cameras, one per month, beginning in June 1959.

The specifications for the camera were:

- 24-inch focal length lens, f/5 lens

- panoramic capability, 35 degrees either side of the vertical

- film transport system to handle 3500 feet of 70mm film

- camera weight not to exceed 200 pounds

- system resolution, not less than 100 lines/mm

In addition, the system had to pass an environmental test to make sure it would work after launch vibration and in a vacuum 150 miles above the earth.

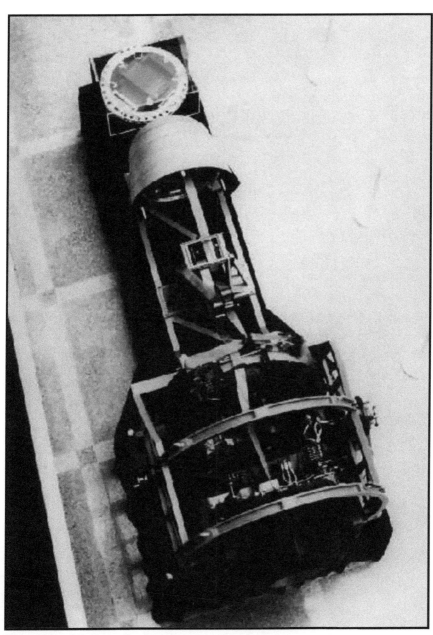

Top View of Corona Camera

Shades of Hycon: I started a six-day, 12-hour-a-day workweek. But unlike Hycon, I convinced management to pay everyone for overtime, including me. This project was a challenge, not only to energize the Itek scientists and engineers to work those hours — all of whom had recently been employed by a university — but also to coordinate the design with Lockheed, GE, Kodak and the customer. There were project meetings twice a month in Boston, Philadelphia (GE), Rochester (Kodak), San Mateo (Lockheed), or Washington.

At Itek, we completed the construction of the first camera in June 1959. After many failures on the vibration machines, we finally passed all the environmental tests in November and began construction of ten cameras.

During this period, we furnished our home and were enjoying the New England countryside. We were also close to our families in New York. Soon after the first camera was completed, I met with Dick Leghorn, now the Chairman of the Board, and Jack Carter, the new president of Itek. They congratulated me and told me that they had a promotion for me.

They decided to open a West Coast office in Palo Alto and wanted me to manage it. It would be a promotion with a big increase in salary. I would continue to manage the camera program since all the cameras would be shipped to the new Lockheed Skunk Works, located in a Hiller Aircraft facility in East Palo Alto. The company's contract included providing a field service team to integrate the camera into the Lockheed satellite. The satellite would be launched from Vandenberg Air Force Base. Several days later the film capsule would be ejected and recovered over the Pacific Ocean by an Air Force C-119 airplane.

I would be responsible to secure and manage additional business, reporting directly to Jack Carter. A building had already been rented in the Stanford Industrial Park. If I accepted, I would be expected to move, open the office and start hiring people soon after the first of the year (seven weeks notice!). Molly and I thought a lot about the offer. We loved our home but had never felt welcomed in Lexington. There was always the question: "Where did you come from and when did your ancestors come to the United States?" Both Molly and I thought that we would like to go back to California, particularly the Bay Area. I accepted the offer. We sold our house within a week.

All of our belongings were packed for us and shipped, including our car. We flew to San Francisco just before the New Year and settled in a motel in Palo Alto. My Lockheed associates introduced us to their housing director, who helped us find a house for rent in Los Altos Hills. It was a few miles up Page Mill Road from El Camino on ten acres. There was a stable with two horses and a dog. Our rent was quite reasonable because we agreed to take care of the animals. Debbie and Victor were told that they could ride the horses. A United Airlines pilot owned the house, and he intended to retire and move back in July 1961.

We jumped at the opportunity. We became friendly with our neighbors. Unlike the Boston area, people didn't ask when our ancestors arrived in the United States. We were on a hill, and below us was a small farm, where Dr. Stevens, the Palo Alto school superintendent, and his family lived. He had a daughter and son the same ages as Debbie and Victor. They became good friends and were in the same classes. I also received a call from a former CCNY physics professor of mine, Professor Simon Sonkin, who was teaching at Stanford. He and his wife held a welcome party for us, where we met many people with whom we became longtime friends.

Back to work. I had to quickly hire a field service crew. I contacted the CIA security people and found out that the U-2 program had been severely cut back, and many of the field service people whom I had hired at Hycon were immediately available.

They could also be cleared very quickly because of their background. A crew was in place by the end of January.

From February until early August 1960, there were eleven launches, five with cameras on board. All failed for one reason or another.

Finally on August 18, there was success. Everything worked as planned. The capsule was recovered over the Pacific Ocean and flown to Rochester where the film was processed. All the project managers and CIA photo interpreters were gathered to look at the results. A ground resolution of 25 feet had been achieved (35 feet was the preflight estimate). The mission covered approximately one million square miles of Russian territory, producing 1400 photos. There were four more flights throughout 1961. The Corona project was a success and became the forerunner of many covert reconnaissance satellite programs.

In June 1960, we decided to buy a lot in Los Altos Hills with a great view of the Bay and build a house. The most important consideration was to keep the kids in their same schools. The house located at 26390 Anacapa Dr. was completed in June 1961, and I lived in it until March 2011.

Itek Palo Alto, as we called it, was growing. We specialized in aerial surveys and photo interpretation. Itek in Lexington, meanwhile, was in competition for the next satellite program but lost to Eastman Kodak and Perkin-Elmer. This, coupled with some ill-advised acquisitions and tumbling stock prices, created a financial crisis for the company. As a result, in February 1962, the Itek Board fired Jack Carter and Dick Leghorn. Franklyn Lindsay, a CIA consultant, became President and CEO.

During this turmoil, Lt. General Donald Putt contacted me. He had been Director of R&D for the Air Force, had recently retired, moved to Atherton, and became affiliated with a Palo Alto venture capital firm, Draper, Gaither & Anderson. I first met General

Putt during U-2 days, when he attended project briefings. He introduced me to this group, who were very anxious to start a company in the Bay Area specializing in the reconnaissance field. They encouraged me to prepare a business plan. I told them that I would consider it only in the event I left Itek.

On May 1, 1962, Mr. Lindsay called me to meet him in Lexington. We met for dinner at his house, where I met his wife and mother. The purpose of the meeting was to inform me that Itek Palo Alto would be shut down and all projects would be moved to Lexington. I was also to return to Lexington. During dinner his mother asked me where I was born, where did my ancestors come from, when did they come to the United States and what schools did I go to. I told her: "I was born in the Bronx, I am a first-generation American, my mother came from Poland in 1908, my father came from Romania in 1912, and I went to City College of New York and Brooklyn Poly."

She told me that Franklyn went to Stanford and Harvard. The family came to the United States in the 1770s. They lived on the Upper East Side of New York but left during the Depression for Carmel because she and her husband didn't want Franklyn to see all those bread lines. That did it for me!

After dinner I told Mr. Lindsay that I was going to leave Itek and start my own company. He asked me to think about it overnight. He said that he would recommend to the Board that I be promoted to vice president when I came back to Lexington. I called Molly that night, and we agreed that I should leave Itek. The next morning I handed in my resignation.

Mr. Lindsay tore it up and said, "It's too late. You're fired!"

When I returned to Palo Alto and cleaned out my desk, I found Itek Palo Alto in a state of shock. Frank Lindsay had called and told the senior staff that I had been fired and that he was in charge. Further, he would be coming out to prepare a plan to shut down the facility in two months. I told several of my associates of my discussions with General Putt. Three of them said that they were ready to quit and help me write a business plan.

I wanted to take some time off before working on a plan. In the meantime, I called Dr. Mees, Director of Research at Kodak. He had told me when I first joined Itek that whenever I left to start my own company, Kodak would help by licensing it with a new product. This would be Kodak's way of getting even with Dick Leghorn when Leghorn and several of Kodak's officers had left to start Itek. Kodak's management was of the opinion that Itek stood for: "I took Eastman Kodak." However, Dick Leghorn claimed it stood for: "Information Technology." Dr. Mees invited me to come to Rochester once I got the new company off and running.

The following week, I received a phone call from Walt Lloyd, Richard Bissell's security chief. He told me that Mr. Bissell and he had been informed by Frank Lindsay that I had left Itek. Mr. Bissell had asked Walt to invite me to come to Washington as soon as possible to discuss my future.

I flew to Washington the following week and met with both Bissell and Lloyd. Mr. Bissell told me that Frank Lindsay had come to see him. Lindsay told him that I had left Itek to start my own company, and I would probably hire away many key people from Itek. He went on to tell Mr. Bissell that the intelligence community considered Itek to be a national treasure, which Bernie Marcus was bent on destroying. He requested the CIA to declare me a security risk and remove all my security clearances from both the CIA and Defense Department. Walt then told me that Mr. Bissell wanted to hear my side of the story.

Epilog: On August 18, 1985, at a ceremony at CIA headquarters in Langley, Virginia, a group of us from the various contractors, the CIA, and Air Force responsible for the first successful Corona mission received the Space Pioneer medal presented to us by the CIA Director, William Casey, and Vice President Herbert Walker Bush. The medal was to be kept secret. On May 18, 1995, President Clinton declassified the Corona project in a ceremony at the Smithsonian Air and Space Museum, and we were allowed to display the medal.

CIA Director William Casey presenting The Presidential Space Pioneer medal, August 18, 1985

Photograph of me and the Corona Camera, Smithsonian Air and Space Museum, May 22, 1995

Only in America!

The Cuban Missile Crisis

Prologue: On July 1, 2015, President Barack Obama announced the resumption of diplomatic relations with Cuba after 54 years.

When I met with Walt and Mr. Bissell in Washington, I told them that Mr. Lindsay had decided that he was going to close down the operation in Palo Alto, which I headed, and that I was to move back to the Boston area where I would await a new assignment. I then said that I would rather resign, stay in the Bay Area where we had just built a new home, and start my own company, which I had always dreamed of doing.

Mr. Bissell told me that he had reminded Mr. Lindsay of how the CIA helped start Itek. Richard Leghorn, a Vice President of Eastman Kodak Company, had resigned and been hired as a consultant by the CIA to get the Corona satellite program going. Then, with the help of the Rockefellers, he founded Itek Corporation and was given a contract to develop and build the cameras for the satellite. He told Mr. Lindsay that the CIA would not revoke my clearances.

Mr. Bissell then offered to hire me as a consultant to start and manage an urgent new project. In addition, if my new company were underway within a month, we would receive a contract for the project. It had to do with the Cuban Missile Crisis. I accepted.

Mr. Bissell told me that President Kennedy wanted to view U-2 stereo photos taken over Cuba. He wanted to tell the world that he had seen the Russian missiles personally. The problem was that he couldn't use a stereoscope to view the photos in three dimensions because of a vision problem. Dr. Land of Polaroid and Dr. Killian, President of MIT and also the President's Science Advisor, had come up with a concept of projecting a stereo image into space like a hologram. A person could stand in front of the viewer and see a three-dimensional image. They already had the optics designed.

I went to Boston to be briefed by Drs. Land and Killian and the optics designer. Then I visited both Kodak and Bausch & Lomb to determine which company could produce the optics in two months and to authorize that company to proceed. I was also to discuss the mechanical design and construction of the viewer to be installed in the White House by September 15, 1962.

After my visit to Boston, I went to Rochester and selected Bausch & Lomb to provide the optical system. Neither Kodak nor Bausch & Lomb would commit to providing a completed viewer in that time frame. I came back to the Bay Area and, together with four former Itek engineers, started Mark Systems, Inc.

We convinced Mr. Bissell that we could and would provide a suitable viewer in the time required. The CIA gave us a contract and a generous advance. We worked with a local machine shop and delivered the viewer to the National Photo Interpretation Center in downtown Washington, D.C., only one week late. There, it was checked out and transported to the White House Situation Room. Several days later, we received word that the President had looked at some images and was delighted.

In mid-October, President Kennedy announced to the nation that he had seen photos of Russian missiles being assembled in Cuba. He warned the Russians to cease and remove the missiles or the United States would invade Cuba and destroy the missiles. The rest is in the history books. That event launched Mark Systems.

A U-2 reconnaissance photo released by the White House to show concrete evidence of Russian missile transporters in Cuba

Epilog: In 1963, Arthur Lundahl, Director of the National Photo-Interpretation Center, told me the following story:

After President Kennedy was assassinated and Lyndon Johnson became President, Mr. Lundahl was briefing the President and his National Security Team showing U-2 photos using the viewer. President Johnson said he wouldn't stand in front of that "G_d damned thing" and told Lundahl to "Get that fxxxxxx thing out of the White House."

Mark Systems, Inc.

A Financial Roller Coaster Ride or All's Well That Ends Well

Mark Systems was incorporated in July 1962 and set up business in Sunnyvale. The venture capital firm Draper, Gaither, & Anderson (DG&A) invested $300,000 for an 80 percent ownership; the other two founders of the company and I invested $25,000 each. In addition, we, the founders, had a significant number of stock options, and we set aside stock options for employees. However, even when all the stock and options were added up, the venture capital group still had a 51 percent ownership and control.

In addition to the CIA, Eastman Kodak Company helped get us started by giving us an exclusive license to develop film processing equipment for the U.S. government using Bimat film, Kodak's answer to Polaroid film. By 1965, our business had grown to $3.5 million in sales. We then moved into a brand new building in Cupertino. Since most of our business was cost plus fixed fee, we were paid monthly and made a profit.

Professor Luis Alvarez now entered the scene. At that time he was head of the Radiation Lab at UC Berkeley. He was well known for his contributions during World War II in helping develop Radar in England and for his invention of GCA (Ground Control Approach) Radar for aircraft. Dr. Alvarez and William Humphrey had applied for a patent for a hand-held stabilized binocular. Dr. Alvarez was very close to Draper, Gaither & Anderson, and our law firm, Cooley, Gaither, Godward, Castro & Huddleson. I had also met him several times when he was a director of Vidya, an Itek subsidiary.

We all convinced Dr. Alvarez to enter into an agreement with Mark Systems to have the company develop and produce the binocular for both military and commercial use.

Since Mark Systems needed more financial and technical resources for such an undertaking, part of the deal was for the company to raise an additional one million dollars. It turned out to be pretty easy. The investors were Bessemer Securities, $500,000; the Bank of America, $300,000; and William (Bill) Witter of Wm. D. Witter Inc., $200,000. Pete Bancroft of Bessemer and Bill Witter were elected to the Board of the company. Our Board of Directors now consisted of General Anderson and Donald Lucas of DG&A, Pete Bancroft, Bill Witter, Mal Malcomson, a founder and Vice President, and Bernie Marcus, President.

The Bank of America's investment was contingent upon the Company taking out a line of credit of $1 million. I accepted. My daughter Deborah recently reminded me that when I told her about the loan, I made her promise not to tell Molly. It was Molly who warned me, when we started the company, not to get beholden to any bank. I discovered she was right.

In 1967, the Board felt that, with the binoculars about to go into production, it was a great time for the company to go public. We had a stock split, and the way the numbers came out, the founder's stock was valued at 10 cents a share. In December 1967, General Anderson resigned from the Board, citing ill health. I was then voted in as Chairman of the Board. Within a week, Pete Bancroft resigned, citing that he was moving to New York City to manage Bessemer Securities and take over as President of the *Wall Street Journal* — both of which his family owned.

In February 1968, we filed with the SEC for an initial public offering of 150,000 shares. Don Lucas and Bill Witter were adamant that we not dilute the ownership of the stock too much. Therefore, 16,000 of those shares were split equally among the directors for their family and friends. Our law firm sent each officer and director a copy of SEC Rule 144, which specified how we could sell our stock after a public offering. At that time an officer, director or ten-percent owner could only sell one percent of outstanding shares every three months, up to a maximum of 5000 shares for a period of two years after a public offering. An officer or director could sell any number of shares three months after he or she resigned from the company. The stock was offered to the public on March 25, 1968, at $8 a share. After expenses and commissions, the Company raised $1 million. In retrospect, I should have insisted that we raise more money. At the end of the first day of trading, the stock closed at $25 a share. We were elated.

By June 1968, the stock price had risen to $60 a share.

Around the World in 40 Days

In March 1968, Mark Systems, Inc. received an invitation to the Paris Air Show in July where the Mark stabilized binoculars would be on display and in use in a helicopter event.

In addition, my family and I had been invited by the Israel Defense Forces to come to Israel.

Molly reminded me that I had promised her when we moved to New Jersey that someday I would take her for a trip around the world. I made plans and kept my promise. Debbie joined us. Victor stayed at home to go to summer school.

We left in mid July and were going to land in Paris on Bastille Day but couldn't. Charles De Gaulle, then President of France, had closed all the airports and borders because of a big uprising against him. We flew into Frankfurt, and in the evening were informed that the uprising had ended and the borders were now open. Molly wanted to get out of Germany as quickly as possible because of what happened in Salonika, Greece, where the Nazis wiped out the Jews including members of her family. I rented a car, and we drove nonstop into France. We attended the air show, spent several days in Paris, and then flew to Geneva, Switzerland.

In Geneva we rented a car and drove to Lucerne, where we visited, by invitation, the family that owned the Rolex Watch Company. They were related to General Anderson, head of the venture capital company that funded Mark Systems. They had purchased Mark Systems stock at the opening and made a great deal of money by selling the stock as it rose. They reserved and paid for a suite in a beautiful hotel in the mountains for us and also gave Molly and me each a Rolex watch.

We went by train to Rome, where we spent a few days. Then we were on to Athens, followed by Tel Aviv.

General Shapiro and other members of the Israel Defense Forces (IDF) greeted us. Molly descended the airplane stairs first, and when she reached the ground she got on her hands and knees and kissed the ground and cried. I joined her. It was a very emotional experience. We spent several weeks in Israel visiting all the holy places. It was awesome. Most overwhelming was visiting the Western Wall and placing a note to God in a crack in the Wall. One of the events we attended was the graduation of a class at the Israeli Air Force Academy. There we met Moshe Dayan, who was using one of our stabilized binoculars to follow the aerial show that was part of the graduation ceremony. Each graduate was in an airplane performing aerial maneuvers. It was an emotional and gratifying experience.

When we were ready to leave Israel, Debbie told us that she planned on immigrating to Israel and had made arrangements to stay. That was a big surprise. Several months later she came home—life there was too hard.

From Israel Molly and I flew to Singapore, where we visited a phenomenal bird park, then over Vietnam, where fighting was going on, to Hong Kong. In Hong Kong, we bought some clothes, all made to order in one or two days.

Our next stop was Taipei, Taiwan, where the president of Air America Airlines met us. It is no longer a secret—Air America was owned and operated by the CIA and was headquartered in Taipei. We stayed in The Grand Hotel, owned and operated by Madame Chiang Kai Shek. We were well fed and treated royally.

Then we were on to Tokyo, where to our surprise, people drove British style on the left side of the road. From Tokyo we flew to Hawaii for some rest and relaxation. Finally in late August we returned to San Francisco and home. It was a great trip.

The trip was a beautiful life experience, and one about which we always enjoyed reminiscing. Thank you Molly for making me fulfill my promise.

Only In America!

Returning to Work

Back in the office, I was surprised and disappointed by what I learned. Mark Systems' stock price had gone as high as $85 a share. It was now at $60, and going down. General Anderson, Pete Bancroft and the Bank of America had sold their shares. Don Lucas had transferred his stock to his family, who started selling. I found out several years later through Bill Witter that following the public offering, the DG&A partners and Pete Bancroft started buying Mark Systems stock to push the price up. They then sold all their stock and made a great deal of money. At the height, my stock was worth over $8 million, and each of my kids' stock about $1 million. I sold some of my stock in keeping with Rule 144.

Soon after I returned from my trip, the company was offered an opportunity to buy a high-definition microfilm technology developed by Fairchild Camera and Instrument Company. After investigating it, I decided that this would broaden Mark Systems and hopefully stabilize and then boost its stock price. Fairchild received $250,000. We hired the key people involved, moved them to the Bay Area, and started a subsidiary, Microform Data Systems. Inc. Mark Systems was now in the venture capital business. In 1969, we took Microform public and raised over $5 million. By 1972, Mark Systems had sold all of its Microform stock to keep itself afloat as government business kept dwindling. Mark Systems stock was now in the low teens.

Bill Humphrey served as a consultant to the company for many years as we built many thousands of the binoculars for military use. Bill Humphrey also worked with Mark Systems and Bausch & Lomb to develop a consumer binocular.

Luis Alvarez, who had been receiving a percentage of the selling price of the stabilized binoculars, invited Canon to visit and explore an opportunity to produce a consumer model. Since I had known the management of Bausch & Lomb (B&L) and they were a major manufacturer of standard binoculars, I approached them. We ended up entering into an agreement with B&L. They also had an option to acquire Mark Systems. B&L did develop a binocular but would have had to price the binoculars over $500. Therefore, there was no mass market since few people would spend that much money. In 1975, B&L gave up. Our stock price dropped below $10 per share. Fast-forward to 1986, when Canon introduced a stabilized binocular into the consumer market. It utilized the same principle as ours. It was one year after Luis Alvarez's patent had expired. Ruthie bought me a pair for $850.

I felt we had to do something different. A friend advised getting into the outside plant telephone equipment business, particularly where AT&T's subsidiary, Western Electric, was not involved. I spent a good part of 1975 visiting the outside plant departments of the Bell telephone companies and came up with the concept of an airflow transducer that the telephone companies' outside plant managers felt fulfilled a need.

The telephone companies pressurized their underground cable with dry air to keep water out and maintain service. However, leaks developed and cables got wet—thereby interrupting service. Monitoring equipment in a central office would display the pressure at the end of a cable that usually ran several miles. A drop in pressure would indicate a leak. To determine where the leak in the cable was located, a technician would go from manhole to manhole to find the leak. An airflow transducer installed on a cable in every manhole, generally 500 feet apart, would help pinpoint the location of the leak. It was, and still is, a very large market all over the world.

In 1976, we developed an airflow transducer and applied for a patent. We began field trials in Northwestern Bell Telephone Company. However, extensive field trials were required in each of AT&T's twenty-two operating telephone companies, which would take years. It was now 1977. Mark Systems was down to about 20 employees from a high of over 100 in 1970. Our sales were under $500,000 per year, down from over $15 million in 1970. We were losing money. In 1970, the Libyans had ordered 1000 binoculars, and in early 1977 they ordered an additional 100 at $6000 each. Payment was to be made upon shipment.

We owed the Bank of America $300,000 on our line of credit. On July 15, the Bank of America called their loan, payable in two weeks. I felt the telephone business was a great opportunity but couldn't raise a cent from any of our previous investors, each of whom had made a great deal of money when the company went public. Our Board voted that the company file for bankruptcy. On August 5, I filed with the clerk of the Bankruptcy Court of San Francisco. It cost $200 cash out of my own pocket, no company checks accepted. Molly was with me; I told her that she was right.

Several weeks later, the Libyan government deposited the $600,000 for the binoculars they purchased. The Bankruptcy Court paid the Bank of America. Microform Data Systems paid the Court $100,000 for the inventory and the rights for the flow transducers, which they placed into a new wholly owned subsidiary named Mark Telephone Products, which I later joined as President. All of the company's creditors received 50 cents on each dollar owed them. Mark Systems was no more, and I started a new career designing and building equipment for use in manholes, a long way from outer space.

In retrospect, what would I have done differently? I probably should have insisted that we raise a great deal more money in our initial public offering. I was not savvy enough about financial maneuvering. I'm sure I joined the ranks of many technical types like myself—naive and trusting.

Despite all the changes and losses, "All's well that ends well." I learned the financial end of a company the hard way. Although I took a bath, I had enough money to proceed with my new career with Mark Telephone Products. The important message I have for my family is: "Never give up!"

This portrait (26"x32") was presented to me by the Bank of America after we announced our intent to have a public offering in December 1967. Ten years later, the Bank did not ask for the portrait to be returned.

Only in America!

Pictures Of The Moon

As part of the pre-Apollo preparations to land a man on the moon, NASA created the Ranger series of rocket missions to take high-quality pictures of the moon and transmit them back to Earth in real time. These images were not only to help select landing sites for the future Apollo missions but also to be used for scientific study. The rockets carried television cameras on board and began transmitting images to the Jet Propulsion Laboratories (JPL) minutes before they crash-landed on the moon.

Ranger 1 was launched in August 1961 and failed. Rangers 2 through 6 also failed. Finally, there was a successful flight on July 24, 1964. Ranger 7's TV cameras started rolling on July 31 and transmitted over 4000 images back to Earth before the rocket impacted on the moon.

About three or four days after the mission, NASA released the photos of the moon to the press. The following week I was watching the Danny Kaye comedy show on TV. He talked about the remarkable feat of sending a rocket to the moon and the ability to transmit all those pictures.

Then he said: "Why did it take so long for NASA to show the world the pictures of the lunar surface? I visited JPL in Pasadena and found out that they had to send the film to a local drug store to get processed, and it took three days. Rocket scientists don't know how to develop film."

To help Mark Systems get started, Eastman Kodak had given us exclusive rights to use their Bimat process for government applications. The Bimat process was developed by Kodak for a reconnaissance spy satellite to process film in flight and have the film scanned and transmitted to Earth. To explain the Bimat process simply, it was a continuous "Polaroid-like" process that provided high-quality positive images in 60 seconds. The Bimat film supplied by Kodak was up to 1000 feet long in a variety of widths.

After I heard Danny Kaye, I called one of our directors, Donald Lucas, whose brother Richard was a deputy director of JPL. I told him the Danny Kaye story and that with the Bimat process we could process JPL's film in near real time. Several days later I received a call from Richard Lucas, who invited me to make a presentation to the JPL Ranger staff. I had a demonstration processor and showed them how the process worked. JPL gave us a contract to build them a 35mm x 1000ft Bimat processor and to provide them with ten rolls of film. We delivered in three months, enough time before the next mission scheduled for February.

An image of the lunar surface from the Ranger 8 mission

February 20, 1965, Ranger 8 successfully sent over 7000 high-quality images back to Earth before its impact. The images recorded on 35-mm film were instantly processed on the Bimat film as they arrived at JPL and projected onto a large screen. I attended the viewing along with the press and Danny Kaye. The following weekend in his regular TV show Danny Kaye described his experience at JPL.

He concluded by saying: "They finally learned how to process film."

A Brief Look Into The Nixon White House

I had two different experiences with the Nixon White House. The first occurred in July 1969. The Apollo spacecraft had landed on the moon. Soon afterward, at Mark Systems, I received a call from Navy Captain Bill Holcomb. He was an aide to the Chief of Naval Operations and moved among the Navy, the White House and the CIA. He was a proponent of the Navy using our stabilized binoculars but was having a difficult time getting them approved. The stabilized binoculars were 10 and 20 power and utilized a patented method of stabilizing an optical image in space in the presence of shake and vibration. When looking through the binoculars, "The world stood still." Bill Holcomb was born and raised in South Carolina and was a REAL Southerner who called me Bernie Boy.

"Bernie Boy," he said, "You will be receiving a phone call from the White House around noon eastern time tomorrow. The President wants to use your stabilized binoculars. This could make your company." When the call came, I felt like I should stand at attention. The caller was Charles "Chuck" Colson.

The phone call went as follows: "Hi, I'm Chuck Colson calling from the White House. I'm an aide to the President. He has heard all about your binoculars from Capt. Holcomb. He would like to use them aboard an aircraft carrier in the Pacific to watch the reentry capsule from the Apollo spacecraft as it lands in the ocean. Can you ship one today by Federal Express?"

"Yes," I said, very much in awe of what was happening.

"Just ship them to me, Charles Colson, The White House, for delivery by 10 a.m.," he said.

Then I asked, "To whom do I send an invoice?"

"What kind of invoice?" he asked. "The President of the United States wants your binoculars. How much are they?"

I told him "$4000," and that we were a small company making a few at a time.

He shouted, "Just ship them and take the $4000 off your god-damn income tax."

We shipped them. Several days later on TV we proudly saw President Nixon aboard an aircraft carrier using our binoculars while watching the re-entry of the Apollo astronauts. About three months later we received our first order from the CIA for 500 binoculars to be delivered to the U.S. Navy. A year later, we received an order for 20 binoculars from the Secret Service with instructions to ship 19, and send an invoice for 20. Thank you, President Nixon.

Mark Systems had many contracts with the CIA to provide specialized photographic and optical equipment. One of the CIA departments with which we did business was located in the center of the U.S. Naval Hospital complex in Washington, D.C., within walking distance of the Watergate Hotel. Now that it's out in the open, I can report that it was called the "Dirty Tricks" department.

Whenever I visited there for contract reviews or to arrange for new contracts, I would spend a week in Washington and always stay at the Watergate Hotel. One of these visits took place in mid-June 1972. I was in the office of one of the CIA officials discussing one of our contracts when his secretary knocked and entered. She told him that the two men from the White House he was expecting had arrived. He asked if I would excuse him and wait in his secretary's office because he had to collect some special equipment for these men. As we changed places, he introduced me. "Bernie Marcus," he said, "Meet Howard Hunt and Gordon Liddy." We shook hands.

In about half an hour, they walked out with a box. As the world found out several weeks later, these two men were the master minds behind the break-in of the Democratic Party headquarters office in the Watergate Hotel orchestrated by the Nixon White House.

This took place one night during the week I stayed at the Watergate.

I became interested in following the lives of these three Nixon aides. Chuck Colson, who was Nixon's private attorney, went to prison. While in prison, he became a born-again Christian. Since his release he has been active in prison reform. Howard Hunt went to prison. Since his release, he has been writing spy stories. J. Gordon Liddy also went to prison. After his release he appeared in a movie and has been a radio commentator.

Through all of this, I was an innocent bystander.

More Stabilized Binocular Stories

There were two interesting sequels to President Nixon's use of our stabilized binoculars. He was shown on television using the binoculars aboard an aircraft carrier as he watched the astronauts splash down in the Pacific in July 1969.

I received a call from the Iranian Ambassador to the United States. He said that the Shah had seen the President on the carrier and had called him to congratulate the United States on the event. He was curious about the binoculars. The following week the ambassador received a call from Mr. Colson of the White House staff, who gave him my name and phone number. He said that he wanted to buy the binoculars and could he have a pair in three weeks. Also, that he would be in San Francisco the following week and would bring a check and a plaque he would like us to attach to the binoculars. I gave him the price, $4000, and told him that he could have delivery as requested.

The ambassador arrived at our plant the following week with a personal check and a gold plaque inscribed:

> "To His Royal Highness
> Shahanshah
> Reza Pahlavi"
> (With the ambassador's name underneath)

I was told that *Shahanshah* was King of Kings.

I informed the ambassador that I had a trip to Washington scheduled and would personally deliver the binoculars in two weeks. He invited me to have lunch with him then at the Iranian Embassy.

Before the ambassador's visit, I had phoned Walt Lloyd, our security contact at the CIA, and told him about this order. He asked if I could stop by CIA Headquarters two or three days before I was to deliver the binoculars and bring them with me. He wanted to set up a meeting with Richard Helms, the director of the Agency.

Soon after my arrival in Washington, we met in Director Helms' office. There were several people in attendance. The purpose of the meeting was to discuss placing a transmitter inside the binoculars. The Director listened and decided against the plan on the grounds that the Shah was a very good friend of the United States. He did not wish to upset that relationship. Several years later, Mr. Helms was appointed U.S. Ambassador to Iran.

The following day I went to the Iranian Embassy to deliver the binoculars. The Embassy, which has since been turned into a mosque, was an imposing structure. When I was ushered into the waiting room, I stepped onto a very posh Persian rug and sat down in an all-encompassing chair. A tall husky guard blocked the door. A very large painting of the Shah stared down at me. I was intimidated.

The Ambassador soon came to welcome me. I handed him the binoculars, which he said he would deliver personally to the Shah as a gift the following week. We then went to the dining room where I met several members of his staff. We were served very fine wines and Beluga caviar and crackers. I clearly remember that because the ambassador made a point of the Iranian and Russian fishermen trying to outdo each other in the Caspian Sea searching for the roe.

During a fabulous lunch, he told me that he had checked on my background and knew that I was Jewish. He informed the group that at the United Nations in 1947, Iran made the first motion to form the State of Israel as a Jewish homeland with the United States seconding the motion. How times have changed!

It was the evening of December 23, 1969, and I was at home. Earlier in the day, I had received a call from U.S. Navy Captain Bill

Holcomb, who told me that he had given my home phone number to a former CIA agent to be known as Agent "X" who was in Libya. As you may recall, Bill Holcomb was the Chief of Naval Operations liaison for the ultra covert CIA operations. Agent "X" had told Captain Holcomb that Col. Kadafy had seen President Nixon with the stabilized binoculars and that he wanted to buy them for his military. At about 11 p.m., Agent "X" called me. He told me that he was calling from Col. Kadafy's palace and that the Libyans wanted to buy 1000 binoculars, and the president of the company must come to Libya and negotiate a contract before the end of the year.

I told him, "No way! Libya kicked out all its Jews and they would never let me in, especially since I have an Israeli stamp in my passport." He said that he was calling from the palace, and that he would fix things up for me. I told him "I don't trust the Libyans and once I land, will I ever get out?" I told him to make excuses for me and that I would send an executive vice president of the company. He replied, "It will have to be before the end of the year, or no deal."

I called my contact in the CIA, Walter Lloyd, the day after Christmas and related the conversation. He replied that they knew about the American in Libya, that it was OK with them as long as I received State Department approval, and that the Agency would help if needed. I then called my Israeli Army contact in New York. His response was that they were very happy the Libyans would have this capability; the Libyans would see how well prepared the Israelis were and therefore would stay in their own country.

It took more than a year to enter into a contract with a Libyan middleman, to get State Department approval, to receive approval of an Arabic instruction booklet and to negotiate 50 percent down and 50 percent upon shipment. The U.S. price for the binoculars was $4000 each. The price I quoted to the Libyans was

$5000. The Libyan said to charge them $8000 each. Mark Systems would receive $5000, and the Libyans would distribute the other $3000 "accordingly," including to Kadafy. We shipped the binoculars nine months later. To this day, I have no idea if the Libyans ever used them.

Not Just Another Stabilized Binocular Story

One experience with stabilized binoculars that is hard to write about was with the President of Mexico.

It was early June 1971. One of Mark Systems directors was Donald Lucas. He was married to a Costa Rican lady, whose uncle was a former President of Costa Rica—Jose Figueres Ferrer. His people fondly called him "Don Pepe." Since Don Pepe was a Mark Systems stockholder, he knew of our binoculars. He told Mr. Lucas that he was going to be in Mexico City visiting the Mexican President—Luis Echeverria Alvarez—and would like to show them to him. The Mexicans had recently discovered oil in the Gulf of Mexico, and he thought their navy should have these binoculars while patrolling the sea near their offshore oil rigs. Mr. Lucas thought that he and I should meet Don Pepe to demonstrate them and that we would come back with an order. Don Pepe agreed.

We met in Mexico City the night before our appointment, scheduled for 4 p.m. on June 10. At 3 p.m. we were picked up at our hotel by a limousine, driven to the Presidential Palace, and ushered into a large, opulent office overlooking the city. Don Pepe remarked that there seemed to be an inordinate number of military personnel stationed around the palace. We sat in the office— 4 p.m., 5 p.m., 6 p.m. came and went. The President's secretary kept telling us to be patient. He told us that the President had some urgent business to attend to and had asked that we wait for him.

It was just before 7 p.m. when President Echeverria finally arrived with several bodyguards. He and Don Pepe embraced, and in Spanish, he apologized for keeping us waiting. He seemed impatient and wanted to get on with the demonstration. I described the binoculars, and he took them. In order to effectively demonstrate them, I would usually hold on to a person's shoulders and shake him while he pushed the activation button to see how the "world stands still." I told that to Don Pepe, who explained it to President Echeverria. I was told to go ahead and gently move his shoulders back and forth. The moment I took hold of his shoulders, the bodyguards stationed in the back of the room came running and pushed me to the floor. The President quickly told them it was OK, and they helped me up.

He seemed impressed with the binoculars but also seemed anxious for us to leave. After a short conversation between Don Pepe and Echeverria, we were ushered out of the office and escorted back to our hotel. Later that night, while watching television after dinner, much to my horror, I learned that President Echeverria had ordered Mexican troops to break up a student demonstration at the University of Mexico campus. The students were objecting to an increase in tuition ordered by the president. At approximately 4 p.m. that afternoon, 25 students had been killed and several hundred wounded. President Echeverria was on the campus at that time, the time scheduled for our appointment.

Several months later, we received an order from Don Pepe for ten binoculars. We gave him a substantial discount. I have been told he resold them to Mexico at a substantial profit and split his profit with President Echeverria.

Our Homes in the Bay Area

Rain, Rain Go Away

In June 1961 Molly and I bought a lot on Anacapa Drive in the Fremont Hills section of Los Altos Hills. The developer of Fremont Hills, Mr. Harry Fitzpatrick, told us the size of the lot was about 1-1/3 acres with a 40- to 50-foot pad cut out of the hillside large enough for a 3000-square-foot house and a backyard. The lot had a great view and was affordable — $8000. We had an architect design the house. Mr. Fitzpatrick's company, Fremont Hills Development Company, built it for us. In 1961, moving into Los Altos Hills was cheaper than buying an Eichler in Palo Alto. This photo was taken from the road above the house at 26390 Anacapa Drive in February 1962. It snowed the previous night.

Our new home In Los Altos Hills

That spring we planted apricot, almond, orange and lemon trees and landscaped extensively.

The years that followed held both happy and sad times. There were some great parties, Victor's Bar Mitzvah in 1963, Debbie's departure for UC Berkeley in 1967, Victor's departure for UC Davis in 1968, Debbie's wedding in 1969. Then the big blow —

Molly's diagnosis with breast cancer that same year. We thought that Molly had conquered the disease, and we took many trips, mostly to the East Coast to visit family.

Twenty years later in 1981- 1982, there were very heavy rainfalls. In December 1982 the storms began again. It seemed like it rained every day.

In January 1983, I covered the hillside with roll after roll of plastic tarp to keep the soil from running off. In March 1983, our hillside collapsed.

Hillside covered in plastic tarp

In February, Ruthie came to visit us from Newport Beach. When the house started creaking, and the deck started breaking apart, she said, "I'll see you later." Little did she know that 10 years later she would become the lady of that creaky house.

We called in a soils engineering firm to give us a report on the condition of our hillside and our home site. Their report was that the soil was not properly compacted and that the hill would continue to give way if the rains continued. The forecast was rain and more rain. They strongly urged us to evacuate as soon as possible. We temporarily moved into my sister Elaine's house. Just in time.

The photos speak for themselves.

Our backyard disappeared

The patio dropped about 12 feet at the rear of house

And the deck fell

The house had to be moved, saved, or torn down. Molly's cancer had metastasized. I decided that we would move the house, save as much as we could, and rebuild on a safe spot on our property.

The soils engineers recommended that if we wanted to save our house we should call in a house mover to move it to a safe place on the lot. He recommended Kelly Brothers. On March 16, Joe Kelly came to look at the catastrophe and said he would come out the next day (St. Patrick's Day) on two conditions. One, that I give him a check immediately for $17,000 and two, that I provide a barrel of beer for him and his crew. I agreed.

On March 17, the work began. They cut off the garage, jacked up the house, and put wheels underneath it.

The house jacked up on wheels

It seemed remarkable how only four people could do all that work. In two days, they moved the house toward the road about 20 feet away from the slide. These two photos show the front of the house against the hillside and the back of the house and the drop in the ground.

Front of the house and no garage

What do we do now? Luckily, Tom Ford—an investor in Mark Systems and in charge of housing at Stanford University—told me of a professor's condo that had become available. We took it.

It was just what the "doctor ordered" for us. It was nicely furnished. There was a deck with a pleasant view and a swimming pool in the complex. The rainstorms kept coming.

To save the house, we had to have it moved again, about 200 feet toward our neighbor's lot. That's where it stood in all its glory while we wondered what was next. The Bay Area was declared a disaster area, which made us eligible for a federal loan of $50,000. Since we had been required to take out Federal flood insurance in the amount of $180,000 to obtain our mortgage, we filed a claim. The Federal Flood Insurance Agency denied it. The agent who visited our site said that if mud had slid onto our house, we would have received the full amount. Since the mud slide was in the opposite direction, "Sorry, nothing." I contacted Senator Cranston, whose home was in Los Altos Hills. He contacted the Agency; he told me that was the law, but the Agency would refund all of our past premiums. I then filed a claim with our home insurance company. Another "Sorry, an act of God" — therefore not covered by our insurance policy.

Soon thereafter a news item appeared in the *San Jose Mercury News*. A California appellate court had ruled that a homeowner with the type of insurance I had could not be denied coverage by a landslide that was not his fault. I contacted the law firm in San Jose that won the case — Hoge Fenton Jones & Appel — and they agreed to handle my case. The fee was one-third of all the money I collected plus out-of-pocket expenses. They contacted my insurance company, quoting the ruling.

The insurance company started off offering $25,000 and ended with $180,000. I accepted, paid the lawyers, and then hired a soils and structural engineering firm to advise us what to do. They came up with a location for the house and a hillside design with three levels. The house would be placed on twenty concrete/steel columns each about two feet in diameter, some of them fifteen feet deep. I then hired an architect to design a garage and studio and an add-on to our bedroom. We also had a large deck planned because we would end up with no backyard. The estimates for soil removal, moving the house to the new site, reconstruction and replanting were staggering — over $500,000.

A former neighbor, Jim Mueller, turned out to be our lifesaver. Jim worked for the U.S. Geologic Survey in Menlo Park. He called and asked me to come to the Survey offices to see an aerial photo of our hillside taken in 1955 before this area was developed. The photo showed that there was a large swale, which had later been filled in to create our building site. I took the photo to the town hall, showed it to the town engineer and was able to get the records of my site. The first entry was a letter in 1958 to the Fremont Hills Development Company advising them that this lot was unfit as a home site because the soil used as fill had not been properly compacted. They were instructed to remove the soil and properly compact it before any construction. The next item in the file showed that a permit had been granted to the Fremont Hills Development Company to build a house for Mr. and Mrs. Bernard Marcus.

Obviously Mr. Fitzpatrick had disregarded the town orders. I contacted my lawyer, and we agreed to sue the Fremont Hills Construction Company. The company had two insurance companies and both came back with the same response: "The statute of limitations of seven years had expired." Their claim was that I knew about the problem in 1961 when we bought the lot and contracted to have our house built. Our lawyer felt that if we continued our suit the insurance companies would make an offer before the trial date, since they would be sure a jury would favor us.

Time marched on — months! Molly was not doing well and was in and out of the hospital. In early January 1984, a month before the trial date, our lawyers recommended that we suggest to the insurance companies that we hire a retired judge to mediate the dispute. They agreed. The judge, Timothy Davis, heard both sides. About a week before the trial was to begin in May, Mr. Davis called me into his office to bring me up to date. Regarding the insurance companies, he convinced them that the statute of limitations began when I read the Town records. He then told me that they claimed they could not find the policies since they were over 20 years old. He convinced them to look harder or he would contact the Insurance Commissioner. They found the policies.

Finally, he told me that my law firm had informed him that they had a very large conflict-of-interest problem. They would like to be relieved from the case since one of the Fremont Hills insurance companies was their largest client. They would recommend another law firm and bring them up to speed. This would delay any action for at least six months. The judge told them that he would advise me to authorize him to write to the State Bar Association and accuse the firm of gross misrepresentation to a client. He would then sue the law firm on my behalf for $3,000,000 unless they carried the case through to a successful conclusion. Furthermore, my contract with the law firm would be null and void. In its place, I would agree to pay only for their out-of-pocket expenses. The law firm agreed.

The trial was postponed one month while a deal was structured. I would receive all expenses to reshape and replant the hillside, move the house onto a stable foundation, rebuild an attached garage, rehabilitate the house, and build a very large deck. In addition, our rental costs would be covered. The amount paid me would be $370,000. The law firm then submitted an expense report to the judge, which came to $150,000. He reviewed it and allowed them $28,000. Mr. Fitzpatrick was told to give Molly and me $15,000 in cash for the trauma he had caused. He claimed that he was broke and could only raise $5000. We accepted.

In May, the architect/construction firm applied for a permit. The Town Council did not like the location where the house was to be moved. They claimed that it was too close to the road and refused to grant a variance. They wanted time to have the Town soils engineer make his recommendation. One month later, he submitted his report. The only safe location was where our soils engineer had determined to have the foundation located. We reapplied for a building permit. At a meeting, I appealed to the town council to let us get started. The mayor agreed on condition that I sign a letter holding the town harmless. A lawyer friend in the audience advised me to sign it. He said it would never hold up in court. I signed the letter that evening.

It took us two years to get through the Los Altos Hills Town Hall approvals and the insurance companies, to remove the fill, and to perform the terracing.

Reconstruction Finally Begins

Work began in July 1984. Four thousand truckloads of hillside were removed and donated to the Stevens Creek Reservoir. The house was moved onto its new foundation and construction began. The hillside was terraced with drainage on each level.

In July 1985 we were able to come back home to a house 25 feet lower and 80 feet north of where it first stood. Even from the lower elevation, we still had a great view. As we settled in and began to enjoy our new house, the trauma of the last two years began to wear off.

The house moved, rebuilt on to its new foundation

Approaching the house from the walkway

Overlooking the valley from the deck

For Molly, this was a palace compared to the hospital and the condo. What we loved most was our enlarged bedroom with a door to the back of the deck and access to our new mini orchard. We had a pottery studio built on the other side of the garage and a workroom under the deck. Molly enjoyed her new pottery studio. She was very creative. It was all her design, and she loved it.

Our master bedroom

Then on October 17, 1989, the Loma Prieta earthquake struck. The house stood solidly on its foundation, but much of Molly's pottery fell off shelves and was destroyed.

The following two years were very difficult for her; she was in and out of the hospital every few weeks, until she passed away peacefully on January 26, 1992.

After a series of interesting events, Ruthie and I were married on March 28, 1993. Ruthie and Victor got together and planned to modernize and brighten up the front part of the house. These photos show the updates.

Our living room

Our den

Our kitchen

In the years that followed, Ruthie and I had many happy times at the house. Lots of friends and family visited and stayed over. We hosted many holiday parties. We especially enjoyed the fruits and veggies from our trees and gardens and our view.

Ruthie and her rose garden

The view from our deck

The Beginning of the End

It all started in February 2009 when Ruthie said she could not sleep in our bedroom. Something in the air was bothering her. She could only fall asleep in the den on the other side of the house while sitting up in the recliner. This went on for more than a month. On one of our visits to our physician, Dr. Scott Wood, he told us that one of his patients had a similar problem and had called in an environmental firm. They were able to identify and fix moisture and mold problems. We called them, and one of their technicians came out immediately.

He collected air samples from all the rooms, which he analyzed for moisture levels and air quality. There was no moisture or mold anywhere in the house, including the attic. He then went under the house in the crawl space, looking and taking pictures. He soon came out and proudly announced that he was sure that he had found our problem. At a far corner of the house, the duct was not connected to one of the vents that exited in our bedroom. In that vent were several dead rats. About fifteen years earlier, we had installed air conditioning units and new ducts. The end duct had not been properly attached, and these critters had established a home where they lived and died. I called the company that installed the ducts, and they came out and fixed the problem.

In June of that year, we had a termite inspection. Lo and behold, we had a massive termite infestation. The company strongly recommended tenting our house and deck. Preparing for tenting was not an easy job. It required a tremendous amount of packing of food and medicinals. Then we had to stay out of the house for three days. As this was going on, our sprinkler system had to be repaired all over the hillside. Then the faucet in our bathroom started leaking badly. A new one was ordered and delivered, but it was too large. Our old one was a discontinued model. Two months later a second one arrived and was installed. Our trees needed a lot of care, so we had the tree man come over and massively trim most of our trees.

Then one night late in July, there was a pungent odor in our bedroom. Once again, Ruthie had to sleep in the den. The next morning, I looked under the house, and there was a dead rat caught in the rattrap. I called our Terminix technician, who removed the rat and recommended we call the *Rat Man* to rat-proof our house, because he had found another one under our studio. We had the *Man* come over, and he and his helper spent all day plugging holes in our foundation and around the house while cleaning out a few more rat remains. Ruthie and I looked at each other and agreed, "That's it!"

Several years earlier, we had looked at the plans of a new retirement center to be built on Stanford grounds and called the Classic Residence by Hyatt.

"No way," we said at the time, "No way would we ever give up our space, our gardens, our fruit trees and our view for an apartment."

In the meantime, the complex had been built and looked very nice. We called and made an appointment with Maryellen Conner, one of Hyatt's marketing staff, who was extremely helpful. On August 3, she took us on a tour. We looked at plans, received a pitch, had lunch, and put our name on the list for a small two-bedroom apartment, which we were told had the shortest waiting time. We told our kids. Deborah and Victor were enthusiastic; Laurie had a problem with us making such a big change but has since become very supportive. With the prospects of a massive downsizing from 3400 sq. ft to 1250 sq ft., we planned to put our house up for sale a year later in the fall of 2010. If it sold quickly, we would move into the Oak Creek Apartments on Sand Hill Road, about a half mile from the Hyatt, and wait to be called.

In October 2010, we signed an agreement with Alain Pinel Realty to sell our house. As a condition, we were to have our house staged. This meant wallpaper would be removed and the interior repainted a bland white. A great deal of our furniture, pictures, rugs and things were also to be removed and replaced. Ruthie was dead set against the staging because our house already felt very comfortable and inviting. She felt strongly that our view and studio were key selling factors. The realtor insisted that full staging was the only way to sell a house in that market. We capitulated.

While this was going on, we stayed at the Marriott Residence Inn on El Camino in Los Altos. The first open house was scheduled for Saturday, November 13. At about 1:30 p.m., as Ruthie and I were driving to the Stanford Shopping Center to take a leisurely stroll, we received a call from the realtor to come over to our house as soon as possible. There had been a break-in and burglary.

When we arrived, three sheriff cars were at the house. We were devastated at what we found. A large window in the living room off our deck had been smashed; the safe in our office had been pried open, everything was gone; a file drawer was open, our 2009 income tax returns with all the backup material were gone; my closet was open, my medals, watches and my father's purple heart medal were gone; Ruthie's closet was open, all her jewelry was gone. We estimated our loss at approximately $130,000. The burglars knew we weren't at home, they knew where everything of value was located, and the sheriff in charge was convinced it had been an "inside job."

We cleaned up the mess and had open houses the following weekend and the first weekend in December, both cold and rainy weekends — nothing — no bites. We agreed with the realtor to take it off the market until mid-January but to keep the "For Sale" sign on the property.

It was Wednesday, December 29, when I received a call from a realtor that her client would like to see the house the next day at 1:30 in the afternoon. I said, "Sure, my wife and I will leave the key box outside," I said.

"No," she said, "My client wants to meet the owners."

The next morning was sunny, clear and cool. I received a phone call from a second realtor about 8:30 that his client would like to see our house at noon and would prefer that we were there.

"OK," I said.

The 12 o'clock people spent about an hour and said, "Thank you," and left.

At 1:30, the first realtor Sonya and her client Benny arrived. We spent a few minutes talking, and then Ruthie had to leave for an appointment. Sonya and Benny looked around the house for half an hour, and then Benny wanted to walk over the property with me to ask about all the fruit trees. We then walked into the studio, which was cluttered with furniture and boxes. I pointed out the kiln and pottery wheel. He exclaimed that he was taking pottery lessons and asked what our plans were for the kiln and wheel.

"We will probably donate them to a school," I said.

We walked back into the living room. He kept exclaiming about the view. He was Chinese and had been conversing with Sonya in Chinese. I asked him how long he had been in this country. He replied that he came here in 1975 from Taiwan at the age of twenty. He was now living in Rolling Hills in Southern California and was a banker. I told him that I had visited Taiwan in 1968.

"Where did you stay?" he asked.

"At the Grand Hotel in Taipei," I responded.

"You were a big-time government official or CIA," he said.

"No," I replied, "I was invited to Taiwan, as the guest of the President of China Airlines. I visited the air base in the suburbs of Taipei, from which Taiwanese pilots flew the U-2 airplanes over China. I was invited because I had been the project manager for the camera systems for the U-2."

"Oh, my God," he said, "My father worked at that airbase for China Airlines on the U-2 airplanes. What a small world! You saved us from the Communists. Let's negotiate, I want your house."

"No negotiation. We have the house listed at a fair price," I replied.

"OK, if the kiln and pottery wheel are included," Benny said, "I'll pay cash, and you can stay in the house until you move into your new residence."

As we were talking, I received a call from a third realtor who came by with his clients at 3:30 as Benny and Sonya were leaving. The third group made an offer contingent on selling their house, which I declined. Benny's written offer arrived at our realtor's office at 5 p.m. We promptly accepted. It was an interesting day. All the prospective buyers were Chinese. The stars must have been aligned favorably.

We sold our house on December 30, 2010, and moved into the Vi on March 15, 2011. Ruthie visited the house several months later and was devastated by what she saw. Her beautiful house had been completely gutted by the new owner. The trauma caused by all of our renovations, painting, repair, Victor's work, the time, the expense, the inconvenience and all those headaches were all for naught.

On the morning of August 11, our former neighbor Palmer Dyal called and told me that he was sending photos of our house on fire. He had taken them at 2:30 that morning. I told him I would get there as soon as I could.

Fire during the night

This photo was taken that afternoon.

After the fire was out

Luckily no one was living in the house.

Ruthie's first reaction was "Molly didn't like what she saw."

My memoirs instructor, Sylvia Halloran's first reaction was, "The bird."

The fire inspector said that the contractor had stacked a bunch of oily rags in the garage and spontaneous combustion triggered by the heat or a spark from the water heater had set off the fire.

My reaction: THE END

Our Palo Alto Cruise

On February 14, Maryellen informed us that an apartment would be empty shortly. We looked at it the following week and signed up. Some renovations were made. On March 15, 2011, Ruthie and I moved into apartment #200C at the Vi at Palo Alto, formerly the Classic Residence by Hyatt, a continuing care retirement community with an array of services and amenities that the management claims are designed to enhance our lifestyle and promote our freedom and well being.

Ruthie was 86 years old, and I was 88. For a variety of reasons, including our age, we were ready to make the move. Looking back, it's as if it was all planned out for us. In Hebrew there is a word *bashert*, translated — it was meant to be.

We are across the street from the Stanford Shopping Center. There is a great swimming pool, fitness center, and delightful pathways to walk. The only decisions we have to make are: what time do we want to eat and in which dining room, what do we want to eat from an extensive menu, which of the many activities, lectures, and concerts do we want to attend, or do we want to stretch out and do nothing? We feel pampered. The people we have met are intellectually stimulating and very friendly.

So far, so good! Six years later, we have acclimated well to our new surroundings, have met many fascinating people, and have made new friends. The activities offered, the conveniences, the concerts, lectures and the dinners are beyond anything we ever dreamed.

All in all, we are very happy to be here. It feels as if we're on a perpetual cruise without the waves. These two photos show the entrance to our home and the complex where our apartment is located. This is the ultimate cruise.

Entrance to the Vi at 620 Sand Hill Road, Palo Alto, California

This is a view of our complex

Only in America!

Life with Ruthie

What Took Us So Long?

I met my wife Ruthie in Pasadena in September 1955. We were married in March 1993.

My first wife Molly and I met Ruthie and her husband, Irving Howard, after we moved to Pasadena from Eatontown, New Jersey, in July 1955. In September my son Victor started kindergarten unwillingly. On the first day Molly waited for him after school, positive that he would come out crying. Waiting next to her was another mother, Barbara Levy. She and Molly talked, and Barbara was also positive that her daughter Peggy would come out crying. To both mothers' surprise, Victor and Peggy came out holding hands.

The Levys invited us, the newcomers, to a party at their house where we met Ruthie and Irv. Our children were the same ages, and we had many family gatherings together. Molly and Ruthie became very good friends, as did our kids, my Deborah and Victor, and Ruthie's Melody and Neil. Deborah and Melody went to a girls' camp together. We moved to Lexington, Massachusetts, in 1958, and then to the Bay Area in 1960. Ruthie and Molly kept in touch on a regular basis. Once a year, we travelled down to Newport Beach, where Ruthie lived. Ruthie and Irv had been divorced in 1970. Ruthie's third child, Laurie was ten years old at that time. After the divorce, when we would take Ruthie out to dinner, my conversation with her was always very casual and short, that is, "What would you like to eat?"

That same year, Molly had been diagnosed with breast cancer. She fought the disease with a vengeance, but it finally got the upper hand, and she passed away in January 1992. I then immersed myself in my business and traveled constantly, visiting customers. On May 14 (I remember that date well because it is my birthday), while asleep, I was suddenly awakened (I thought) and was startled by Molly standing next to the bed.

"Molly, what are you doing here?" I asked. "You're dead."

"I'm around you all the time," she said, "and I want you to do something for me. Ruthie is in need of help and seems to have no one to turn to. You must go down to Newport Beach and do all you can do for her. If you don't, I shall haunt you for the rest of your life."

Then she disappeared. Those of you who know me realize I never believed in any of this paranormal "stuff" until that moment.

Two weeks later, I arranged for a trip to Southern California, called Ruthie and told her that I was in town and asked if would she like to have dinner with me. At that time I found out what was troubling Ruthie. She was in a state of shock. Melody had been diagnosed with a brain tumor. We took a very long walk on the beach and caught up with our lives. Ruthie couldn't imagine that she walked so far, and I couldn't imagine that I talked so much. It was then that we knew that our relationship was no longer a casual one.

During our walk Ruthie told me that Melody was going to have an MRI the next morning, followed by a report from her doctor. I told Ruthie that I would like to go with her and support her. She called Melody, who agreed. I accompanied Ruthie, Melody and Melody's husband Ralph to hear the results of the test. The doctor indicated that the next step was an invasive operation to determine the type of tumor and how to treat it. She would be hospitalized for several weeks. Ruthie's cousin Merle called her and told her to obtain a second opinion. I agreed and called a doctor who had treated Molly. He told me that UCSF hospital in San Francisco had the best brain tumor clinic in the world and gave me the name of a well-known doctor. We convinced Melody to go. The end result was a very simple procedure and an overnight stay in the hospital. The diagnosis was that the tumor was untreatable but could be contained with the proper medication. Melody lived for 12 more years.

Ruthie and I became engaged in August. Before we were married, I told Ruthie of my vision of Molly and the instructions I was given.

We were going to have a small wedding with just our kids. Melody objected, "What about all your family and mutual friends?" We gave in to her and started planning a "real" wedding. When Ruthie and I began making arrangements, we were asked to bring our daughter, the bride. They smiled when we said, "But it's our wedding!"

A new chapter in our lives began on March 28, 1993, the day we were married, with the blessings of all our children, grandchildren, many relatives and mutual friends totaling 200 guests. Ruthie was 68, and I was 70. It was a beautiful wedding ceremony. Many of our guests to this day comment on the love they felt Ruthie and I shared, as well as the love all around the room. Did Melody or Molly arrange all this, or was it just meant to be?

"I now pronounce you Man and Wife. You may kiss the bride."

The Family. Left to Right: Victor, Deborah, Neil, Laurie, me, Ruthie, Kori, Ryan, Melody, Ralph

Walking down the aisle after we were married

Molly Is Still Around

Several times a week, starting in the spring of 1994, a quail would fly on to our deck, come up to the living room door and look in. Usually it would get Ruthie's attention by pecking at the door.

One day Ruthie called my attention to the bird, "Look, Bernie," she said, "This bird really wants to come into the house. It's been here several times and keeps pecking away and looking at me. I always smile at it. I would like to let it in, but it's a wild bird."

Later that summer, Ruthie's son Neil and his lady friend Leslie were sitting on lounge chairs on the deck when the quail flew up to the living room door and started pecking, not paying attention to the two people sitting nearby.

Ruthie told her son and his friend, "Look, there's the bird I was telling you about."

Acknowledging her, Neil said, "Isn't that something, the bird doesn't mind us being right here."

Ruthie then overheard Leslie say to the bird: "Everything is ok, Molly, you can go away now."

The quail flew off and never came back. Leslie did not know about Ruthie's thoughts about the quail nor about my experience one night in May 1992.

Ruthie was really upset. She lost her best friend again. She loved the idea that the bird kept coming to the house. She had never had an experience like this before. She felt as if someone was trying to send her a message and thought that it might be Molly.

In December 2000, my daughter Debbie's very dear friend Gage Taylor passed away. Gage was a great artist, having won an award as the outstanding artist of the State of California earlier that year. After he died, Debbie started taking lessons and began painting in Gage's style. In the summer of 2001, she decided to attend an event conducted by a famous psychic being held in a field on the shores of the Hudson River in upstate New York. She arrived late. There were several hundred people in attendance. Soon after she sat down, the psychic asked, "Whose mother is Molly?"

Debbie looked around to see if anyone stood up to acknowledge the question. She was the only one to stand up. He then said, "Your mother just told me that she is very sorry about your loss and very happy that you have taken up painting."

Debbie called me that night with a very shaky voice to tell me her story. After she returned to California, I told her my Molly experiences.

I do believe that Molly is watching over us.

Honoree

My name is engraved on the National Aviation and Space Exploration Wall of Honor of The Smithsonian Air and Space Museum for my work on the U-2 and the satellite programs.

The wall recognizes those who have made achievements in flight. The Exploration Wall of Honor is outside the Steven F. Udvar-Hazy Center Museum in Chantilly, Virginia, near Dulles Airport. The Udvar-Hazy Center, opened in 2003, displays many of the aviation and space artifacts that can't be shown on the Mall due to space limitations.

The Wall of Honor is a permanent memorial to the thousands of people who have contributed to the U.S. aviation and space exploration heritage. Names of honorees are inscribed on the airfoil-shaped wall, which will continue to grow in the future.

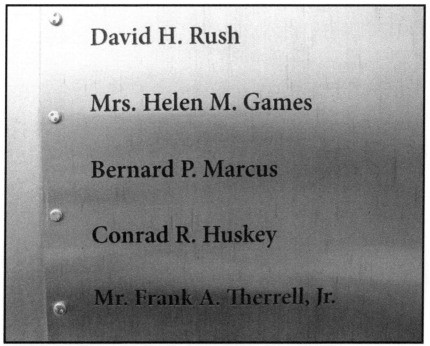

A close-up of the plaque showing my name

The National Aviation and Space Exploration Wall of Honor

It's Never Too Late

In July 1931 when my sister Millie was six years old and I was nine, my mother decided that Millie should take piano lessons. We were living at 440 Jackson Ave. in the Bronx. My father bought a Steinway upright. He was able to get a good deal because he had worked at Steinway from 1912, when he first came to this country, until 1917, when he joined the Army during World War I. He had maintained contacts there. I can still see the piano being hoisted up two stories in the front of the apartment building. One of the living room windows had been removed and the piano was moved into our apartment. I begged my mother to allow me to take lessons with my sister, and her response was "Girls take piano lessons, boys play the violin."

Six months later, a truck arrived and I watched the piano being lowered down the front of the apartment building. Millie did not play the piano, and I did not play the violin.

In July 1961, Molly, our children, Deborah and Victor, and I moved into 26390 Anacapa Drive. Los Altos Hills. Molly decided our living room needed a baby grand piano. We bought a Hamilton baby grand. It was moved through the front door into the living room, where it stood majestically. It was the first thing you saw when you walked into the house.

Molly thought that Deborah should take piano lessons, which she did for about a year and a half, and that was it. I thought about lessons, but I was too busy at work.

Time marched on. Our children left home for college and then their careers. In 1985, when he was eight years old, my grandson Jory, Deborah's son, told me that he wanted a violin. I took him to a musical instrument store in downtown Palo Alto and bought him his first violin. Today he is an outstanding musician, assuring me that he is making a good living. He was trained at the Crowden School, the San Francisco Conservatory of Music and the New England Conservatory of Music. We are all very proud of him. I'm positive my mother is also pleased and proud. Molly died in 1992. Ruthie and I were married in 1993. Victor remodeled the house in 1993. All through this the piano was well dusted, had loads of photos and artifacts on it and remained unplayed.

In 2001, we sold our company, Mark Products. Part of the deal was that I would stay on as a consultant until my 80th birthday, May 14, 2002. Ruthie knew about these piano stories, and she encouraged me to take piano lessons when I retired. I asked a musician friend, Mike Elliott, to refer me to a good teacher. He encouraged me to call Dr. Isaacs, at the time the music director at the Community School of Music and Arts. Dr. Isaacs interviewed me to make sure that I was serious about taking piano lessons at my age. I told him that it would also be a thrill to accompany my grandson. He signed me up, and I began taking lessons soon after my 80th birthday.

When Ruthie and I moved into our retirement community, we sold our baby grand and bought an upright, which fits nicely in our living room. It was moved into our apartment through the front door. I am still taking lessons at the Community School and enjoying them very much. Whenever I perform before an audience, I always play Beethoven's *Fur Elise*.

My grandson Jory and me – circa 2008

Me – January 2013

Rosh Hashanah 5777

Rosh Hashanah 5777 or the Jewish New Year 5777 (lunar calendar) occurred on October 3, 2016. In 1958, Rosh Hashanah 5719 occurred on September 15. What is significant about those dates? Here is the story.

In July 1958 we moved from Pasadena, California, to Lexington, Massachusetts. My job changed from project manager at Hycon for the U-2 camera systems to project manager at Itek for the camera system for the Corona satellite. The CIA ran both programs, which were *highly* classified. Richard Bissell (who wished to be known as Mr. B) was the Deputy Director of the CIA at that time and was in charge of both programs. Other companies involved in the satellite program were Lockheed, GE, Fairchild, and Eastman Kodak. Progress meetings were held once a month, rotating among the cities of all four companies and in Washington at innocuous locations. CIA couriers delivered the meeting dates and locations orally. The Agency, or *"The Company"* as it was called, was concerned about non-secure telephone lines and the public mail.

The September 1958 meeting date was to be on the 15th, and we were to confirm our attendance to the courier. I told the courier that I could not attend on that date or on the 24th. I told him that the 15th was Rosh Hashanah, the Jewish New Year, and the 24th was Yom Kippur, the Day of Atonement. He said he would report it to his superior, Walter Lloyd, the Director of Security. A day or so later I received a phone call from Walt Lloyd.

"Bernie, what's with this Rosh Hashanah and Yom Kippur? Mr. B asked me to find out all about it. You have to show up at the meeting, after all this is the highest priority program in our country, and he put you in charge of the camera systems."

I told Walt what those days symbolized:

180

"Rosh Hashanah is celebrated as the Jewish New Year. It occurs in the early autumn and is the first day of the first month of the Jewish calendar. It is also the first day of the High Holy Days, which end on Yom Kippur ten days later. Rosh Hashanah is believed to be the anniversary of the creation of Adam and Eve. Walt, when I was a kid, I never went to school and, as a grown-up, I've never gone to work on the high holidays. I always go to the temple with my family. Also on Yom Kippur, it is required that one fasts, because on that day God inscribes each person's fate in the Book of Life."

"OK, Bernie," said Walt, "I'll let Mr. B know all this and get back to you."

Several days later Walt called me to tell me that the meeting was now scheduled two weeks after the original date, same place. He also told me that he had been instructed to check on a Jewish calendar. If a meeting was to be scheduled on a Jewish holiday, he should call me and find out if I would attend. If I would not, the meeting date would then be changed.

Over the ensuing years, Walt Lloyd and I and our families became good friends. Walt was stationed in Palo Alto in 1961 and 1962, and Molly, the kids and I would spend Christmas Eve at his house with his family, and they would spend one night of Passover at our house. Walt later became Chief of Security of the CIA, then Chief of Station in Iran, and finally Chief Counsel of the Agency, after which he retired.

Every year since 1958, on Rosh Hashanah, we receive a New Year's greeting from Walt. This year was no exception. On October 3 we received a "Happy New Year" call. We have a very special friendship.

What a great country we live in!

Only in America!

Take Me Out to the Ballgame

I went to my first baseball game when I was ten years old. My older cousins, Nat, 14, and Irv, 16, took me. We rode the 149th Street trolley car to the Grand Concourse and then walked to 161st Street and River Ave. to Yankee Stadium. We sat in the bleachers, of course. Since I was under 13, I paid 25 cents, and my cousins paid 55 cents. It was on May 30, Decoration Day, and there was a doubleheader against the Boston Red Sox. It was very exciting. Babe Ruth was in right field, Lou Gehrig on first base, Bill Dickey was the catcher. I remember that the Yankees won both games. As a matter of fact, I can't remember ever going to a ballgame that the Yankees lost. From then on until I went into the Army, I only went to doubleheaders. It was like going to a double feature at a movie theater — it wasn't worth the money to see only one game. I was a rabid Yankee fan, hated the New York Giants, and always rooted for the Dodgers against the Giants.

After the War, we moved to New Jersey and then to Pasadena in 1955. My baseball season became limited to the radio and then TV — I didn't travel to see a live game. In 1958, the Dodgers moved to Los Angeles. I thought that I would finally be able to go to a ballgame, but we then moved to Lexington, Massachusetts. I couldn't root for the Red Sox or go to see them play. In December 1959, we moved from Lexington to Los Altos Hills, about the same time as the Giants moved to San Francisco. My kids didn't have a favorite team, so I would take them to Candlestick Park to watch the Giants play. Due to its location next to San Francisco Bay, strong cold winds often swirled down into the stadium. We froze and often had to wrap ourselves in blankets or sleeping bags. As time went on, we became Giants fans, and the Dodgers became the enemy.

In the 2000 season, the Giants moved into what is now known as AT&T Park. My longtime friend Dean Mack and I bought season ticket rights for section 110, row 38, seats 3 and 4. They were great seats between home plate and first base in the shade under the overhang. We went to a lot of ballgames. Then Dean moved to Friday Harbor, Washington, in 2006.

Eighty-four home games per season were too many games to go to, or to sell, give away or just throw the tickets away, so we sold our rights just before the 2010 season. To our amazement the Giants won the World Series in 2010. Oh well.

From time to time since then, I have been going to ballgames. One of those days happened to be June 27, 2012. It was a day game. Bernie Arfin, a fellow resident at the Vi and an ardent Giants fan, had tickets to the game — the Giants versus the hated Dodgers. The seats were in section 122, row 26, seats 7 and 8, behind the Giants dugout. The list price for my seat was $68! It was the top of the seventh inning, the Giants were winning 3-0, the Dodgers were up, Lincecum was pitching, and the batter hit a high foul ball. We were all watching as the ball headed our way. It fell on a railing a few rows in front of us, bounced up in the air – then down on my side, landing under my seat – yes, right at my feet. I don't remember who the batter was. It's the first ball I've caught in 80 years of going to ballgames. Bernie and all the fans around me were excited that this "old" Giants fan got the ball. Needless to say, I was thrilled. It now sits on a vase on my piano — prominently displayed.

I caught the baseball

Thoughts on Life

When Irish Eyes are Smiling

St. Patrick's Day and Easter Sunday bring to mind both good and bad memories.

In the 1920s we lived on the ground floor of an apartment house on 138th Street in the Bronx, just up the block from St. Ann's Irish Catholic Church. The neighborhood was mixed Irish and Jewish. I clearly remember that every Easter Sunday my parents would pull down the shades in our apartment, turn off the lights, lock our front door and keep me indoors all day as if they were still in Europe. When services were finished, the Irish kids would run out of the church looking for and yelling for the Jewish kids to come out onto the street so that they can beat them up for "Killing our God." My mother would take me to school on the following Monday morning when things hopefully had quieted down.

When we moved to Jackson Avenue, our new neighborhood was mixed Italian and Jewish, and we all got along fine. When it was time to go to junior high school, which was about a mile away on the other side of St. Mary's Park in a solid Irish neighborhood, my Italian friends and I would walk together for safety reasons. The Irish kids were always looking for a fight calling us "Kikes and Ginnies." We would always respond by calling them "Lousy Micks."

Despite all the name-calling, every year my friends and I went to Fifth Avenue to watch the St. Patrick's Day parade. It was usually cold and rainy, but we cheered the marchers and bands.

On March 28, 1993, change began. Ruthie and I were married. Marching down the aisle as part of the wedding party at the synagogue in Newport Beach were Ruthie's two half-Irish grand-kids, Ryan, 17 years old, and Korrine, 14 years old. Ruthie's daughter, Melody, had been married to Neil Burns, a man of Irish heritage. The kids and I have been close and fond of each other since they were born. Years later, when they were married, I was at the altar delivering both a Hebrew and an Irish prayer for them. In addition, Ruthie's cousin Merle married Mary Bradley, a first generation Irish-American, with whom we are very close, as we are with all of her Irish relatives.

In 2004, I began volunteering for the Brain Injury Unit and the Post-Traumatic Stress Disorder Unit at the VA Hospital in Palo Alto. The VA nurse in charge was Paul Johnson, a nice Irish young man. He was very competent and kind, and we have become good friends. A fellow volunteer and vet was Frank O'Neill, a Boston Irishman. He and I became very good friends as well. Frank and I formed an organization called *Vets Helping Vets*, through which we raised money to treat hospitalized vets to experiences the VA could not afford to give them. For example, we have taken them to Giants and A's baseball games.

Frank is the epitome of a politician. He ran the Kennedy presidential campaign in Boston. He ran the Dianne Feinstein senatorial campaign in San Francisco. He is a retired Port Commissioner for the City and also chaired the Hunter's Point Redevelopment Agency. His uncle was Tip O'Neill, formerly Speaker of the House of Representatives, whom I met in 1973. At that time, Mark Systems needed government approval to obtain a large contract from Libya for our stabilized binoculars. For a slight contribution into a barrel in front of his office marked *The Buck Stops Here*, the Speaker introduced me to the Chairman of the House Armed Services Committee, who arranged for the approval of our contract.

Frank has also been the President of the Hibernian Club, head-quartered in Burlingame. Ruthie and I have attended many of the

St. Patrick's Day celebrations hosted by the Club. We always sat at the head table with Frank and his wife in a dining room with about 200 guests. Also seated at our table were visiting officials from Ireland and Father Andy Johnson, who would deliver the blessings. As he introduced the visitors at the head table, Frank would always introduce his two Jewish friends to the audience. Following the 2012 St. Patrick's Day bash, which we were unable to attend, he told me that following the opening prayer, Father Andy called from the podium to Frank and asked where his two Jewish friends were. Then he said, "Tell them we miss them." Frank also told me that Father Andy had been appointed a Canon of the Church of the Holy Sepulchre in Jerusalem and was all dressed up in red robes with a red yarmulke on his head.

In 2016, Ruthie and I attended the St. Patrick's Day celebration in the auditorium of the Vi at Palo Alto. There was a band on the stage playing and singing Irish music. One of the musicians was Michael Halloran, the husband of my memoirs teacher, Sylvia Halloran, who was also present—a small world. We joined a sing-along of *When Irish Eyes are Smiling*.

Time marches on, and I celebrated my 95th birthday this year. I am so grateful to have lived this long and to know that we of different backgrounds and heritages can be friends and family.

We've Come a Long Way

For my children, grandchildren, great grandchildren, and for my nieces and nephews, I would like to recount my experiences throughout my life with Black Americans.

I was born in the South Bronx in 1922 and lived there until 1942. Segregation of the blacks during those years, even in New York City, was an accepted fact. Saturdays were the days that all the kids in the neighborhood went to the movies. We were not allowed to go up to the second balcony. That was reserved for the blacks who had to enter the theater through the backyard by climbing up a fire escape. It was known as "N (word) Heaven." This period was also the time that Joe Louis became the heavy-weight-boxing champion of the world. After each Joe Louis victory, the black kids in Harlem would feel their oats and try to come across the bridges into the Bronx. We, the white kids, had gangs and would line up on "our side" of the 149th Street bridge with bags of rocks to keep the black kids on their side in Harlem. I started going to City College of New York in 1938. I would take a trolley car on 149th Street across the same bridge into Harlem where the College was located. We had over 20,000 students. Not one was black!

I was drafted into the Army when I graduated from college and was sent by train, with several hundred other draftees from New York City to Camp Wheeler, near Macon, Georgia. We stopped for several hours in Savannah. We, of course, had to go to the men's room and were warned not to go into the "Colored only" facilities nor drink water out of the "Colored only" water fountains, nor sit in the "Colored only" waiting room." It felt like a different, scary world.

During a weekend trip three of us tried to take from Camp Wheeler to Savannah during basic training, we were sworn at and kicked off the bus by an armed police officer because we offered our seats to a pregnant black woman and her child. That was an abrupt lesson to us of the difference between the North and South.

In 1953 Molly, the kids and I were living in Eatontown, New Jersey. Debbie was five years old, and Victor was three. Molly and I wanted to take the kids to the Shrewsbury River to watch boat races. The best viewing spot was in the town of Rumson, situated on the banks of the river.

As we are about to enter the town, there was a gate and a guard and a sign that read: "Negroes, Jews and dogs not allowed."

We turned around.

It was the late 1950s. I was in Washington, D.C., on business and staying in a hotel in Arlington, Virginia. I decided to take a bus at the end of the day instead of a taxi from Washington to Arlington. The two cities are separated by the Potomac River. As the bus came to the midpoint of the bridge, it stopped. Every black person got up from their seats and moved to the back of the bus. Nobody said a word. That was the law in Virginia.

Fast forward to 1968. I was President of Mark Systems. We had several large government contracts. Every month, I had to fill out an Equal Opportunity Survey and mail it in to the federal government. One day, I had a visit from an agent of the Equal Opportunity Commission. He advised me that the company was in violation of the law. We had three blacks on our payroll, and according to the racial mix in Sunnyvale, California, we were only entitled to one. I showed him to the front door and did not hear from the Commission again.

Time marches on. The United States of America elected a black man, Barack Obama, as President. We've come a long way!

Health Care Reform

As the Congress debates health care reform again, I feel compelled to relate my experiences with an insurance company and the Medicare prescription program. This is addressed specifically to my family and friends.

In the autumn of 1991, after 22 years, Molly was having a terrible battle with cancer, which was spreading all over her body. She was undergoing radiation treatments at the Palo Alto Medical Foundation and was in and out of the Stanford Hospital. In September, she was hospitalized to remove water from her lungs, which would take several days. One morning, I received a call from the hospital that our insurance company, Aetna, had notified them that Molly's insurance coverage had reached its lifetime limits. I was asked to come in to assure the hospital that I could pay for her continued care, or I would have to take her home or to a county hospital. I called our Aetna representative, with whom we had been insured through my company since 1962. He said, "Sorry, the fact that your company had been contracting with Aetna since 1962 makes no difference." Thank God that we had enough money to keep providing her medical care, for the many more times she would be treated in the hospital, and for treatments at the Palo Alto Medical Foundation. We were out-of-pocket for a considerable amount of money.

Molly died in the hospital in January 1992. Several weeks later, the company cancelled Aetna and joined a Silicon Valley cooperative.

Today, I'm lucky. I have "socialized" health care. I have Medicare, a Medicare advantage program, and Veterans Administration health care, which includes prescription coverage. Ruthie has Medicare and a Medicare advantage program, which includes prescription coverage as passed by the George Bush administration. However, in the fall of each year, she finds herself in what is called the coverage gap or "doughnut hole." She has been co-paying up to $2700 in total drug costs. In the coverage gap, she must pay full retail for the same drug; for example, she pays $198 per month for Nexium. Through the VA, I pay $8 per month. By law, Medicare cannot negotiate pricing with the drug companies, but the VA can. This coverage gap continues until Ruthie pays an additional $1650 out-of-pocket, and then the Medicare co-pay begins again.

Fortunately, we can afford both food and our medications. There are millions at this time who have to make a choice—food or medicines. I'm for the Congress passing a meaningful health care reform bill. I hope you are too.

To Jory and Natasha,
Two Great Violinists

On Your Wedding Day, June 19, 2016

For several weeks I had been practicing one of my favorite Beethoven pieces, *Fur Elise*, to be played on your wedding day, but unfortunately the piano was unavailable, stuck in a corner. There are many speculations as to who was Elise? Was she a student, a lover? Suffice it to say it is a very beautiful piece.

Jory and Natasha

Now, Natasha, I would like to give you a little history about Jory and me. When I was eight years old, living in an apartment house in the Bronx, my mother decided that my sister Millie, then five, should learn to play the piano. Mom had a piano hoisted into our apartment. I pleaded with her that I wanted to take piano lessons as well. Her response was, "Girls take piano lessons, and boys play the violin." I didn't want to play the violin. Five months later, the piano was hoisted back down the apartment building - Millie didn't want to play the piano.

Time marches on, it is March 1955, and I am sitting before three professors at Brooklyn Polytechnic Institute being quizzed about my thesis for my Masters in Physics. One of the inquisitors was Professor Isador Fankuchen. Twenty-two years later, my grandson, your husband Jory Fankuchen was born. Professor Fankuchen was Jory's great uncle. You must admit this is a very small world. When Jory was eight years old and visiting Molly, his grandmother, and me at our house in Los Altos Hills we asked him what he would like for his birthday, "A violin," he replied. I took him with me to Palo Alto to a musical instrument store, and he picked an "expensive" violin - $10.

Of course he progressed from that. Important memories I have of him playing were:

- On January 25, 1992, Molly was in Stanford Hospital. She had been there for several weeks and was not given much more time to live. Jory came to the hospital to play for Molly. We gathered in a large area of the hospital and Jory played Mendelsohn's violin concerto before a rather large crowd. Molly started crying, and she whispered to me, "Bernie, I am now ready to die." She passed away later that night.

- In the mid 1990s, I can't remember the exact date, the San Francisco Conservatory was celebrating Isaac Stern's birthday. Jory was selected to play a solo for him on Yehudi Menuhin's violin. I was a very proud grandfather.

- Sometime in 2007 Jory visited us at our house to play a duet with me. On May 14, 2002, my 80th birthday, I had started taking piano lessons, which I continue doing to this day.

- On May 14, 2012, my 90th birthday, I met you for the first time. You and Jory had flown in from somewhere to my birthday party and entertained me and my guests with a series of beautiful violin duets. You were great.

I look forward to providing my great grandchildren with the musical instrument of their choice.

My love to you both,

Papoo

Other Memories That Come To Mind

In May 1969 student dissidents were protesting the Vietnam War at universities all over the country. UC Berkeley was no exception. Ronald Reagan was governor of California at the time and had stationed the National Guard in Berkeley. The Guard had thrown tear gas many times at students as well as firing rubber bullets at them. One of those protest days occurred in Peoples Park, and Debbie was one of the protestors. The Guard broke up the protest using tear gas and firing rubber bullets at the students. Debbie was gassed and hit with a rubber bullet in her butt. After that experience, Debbie became a staunch protestor and, needless to say, that further diminished our opinion of Ronald Reagan.

The National Guard at Berkeley

By the mid 1970s, Victor achieved his childhood dream of becoming a proficient mountain climber. He was also a mountain ranger at Yosemite. One day in early summer, he called Molly and me to tell us that he was going to climb Half Dome. He said that he would like us to be there and had made a reservation for us at the Ahwahnee Hotel. We went and watched him climb, following him up the mountain with our stabilized binoculars. About 4 p.m. he stopped climbing. We, and other hundreds of others, watched him take a sling out of his backpack, hook it up on some rocks, wave to us down below and lie down. He spent the night on the side of Half Dome, and we spent the night at the Ahwahnee, petrified. Early the next morning, a ranger called to tell us that Victor had reached the top and all was well.

Victor climbing Half Dome

Our Ancestors

To Deborah, Victor and Jory: I've finally decided to write down all I know about our ancestors. I encourage you to investigate further. Going back in history, there are no written records of Jews who lived in Eastern Europe (i.e., the Russian empire including Poland) prior to Napoleon's conquest, when they were directed to obtain surnames for census taking and taxation. Also, in a 1787 law, the Austro-Hungarian Empire, which included Galicia and Romania at that time, required Jews to register a surname, which gave the government control over these city dwellers in such matters as tax collection and military conscription.

My father's ancestors picked the names Marcus and Altarescu; my mother's side picked Lerner and Gross. On the other hand, your mother Molly's ancestors, who fled from Spain to Turkey during the inquisition in the late 15th century, had surnames at that time and for centuries before. Her father's side was Cohen; her mother's was Crespi.

My father, David, was born in 1896 in Pascani, Romania, which was near the Ukrainian-Russian border. His father's name was Hersch, his mother's was Mangha. They had five children: Sophie, Ethel, Joseph, David and Chai. My grandfather Hersch had a bakery shop. My father told me that there was a hidden trap door to a basement in the bakery where the family hid when they heard people yelling "The Cossacks are coming." The Cossacks would often cross the border into the Jewish ghettos, looking to kill male Jews and abduct young females. In 1910, my father's two older siblings, Sophie and Joseph, decided to leave for the United States. My father followed them in 1912 at the age of 16 and lived with them in a large apartment in The Bronx on 138th Street. He was hired by Steinway Piano to be a piano finisher. In 1917, he joined the Army and went overseas during World War I. He returned in 1919, having been wounded and gassed in combat. After the War, the Army sent him to school to learn a trade, and he chose to become an optician.

My mother, Betty, was born in Premeshlanya, Galicia, in 1900. On that date Galicia was part of the Austro-Hungarian Empire. Today it is part of Poland. Centuries ago it was part of Russia. Her father, Max Lerner, married Rachel Gross in 1899. They had six children. Three were born in Europe: my mother Betty, Harry and Abe. The other three—Sarah, Celia, and Beatrice—were born in the United States. My grandfather was a tailor. He came to this country without his family in 1907 to escape conscription in the army and sent for them the following year. They lived in the lower east side of Manhattan on Rivington Street.

It has constantly amazed me how my grandmother Rachel, a young woman with no education and no reading skills, could pick up, leave a small *stetl* in eastern Poland, travel by horse and cart, then by train to Hamburg, and board a ship to the United States.

My mother and father were married in 1921. We never talked about how they met. They had three children—yours truly, Millie and Elaine. We initially lived in the South Bronx on 138th Street, a block away from my Aunt Sophie.

Now to Molly's side of the family. According to her father, Victor Cohen, his ancestors came from Toledo, Spain. During the inquisition in 1492, his family fled to Portugal, and then on to Salonika. At that time, it was part of the Ottoman Empire and is now a city in Greece. The Turks were very helpful to the Jews at that time and settled them throughout their empire. Molly's mother was Esther Crespi. Her ancestors, also from Toledo, fled to the island of Majorca, and then to Ankara, now capital of Turkey.

Victor Cohen came to this country in 1912 at the age of 19 with his mother Matilda. His father had died when Victor was an infant. They left because they heard of the opportunities in the

United States. He had no job, and they had very little money. They could only afford steerage and knew they would not eat the food on the ship because it was not kosher and was also expensive. Before leaving Salonika they hardboiled several dozen eggs, which was all they ate for ten days on the ship. They settled in the East New York section of Brooklyn, which had a large Sephardic community, and he became a tailor.

Esther Crespi also came to this country in 1912 at the age of 11. She was an orphan at the time, living with an aunt in Ankara. She was terrified when the Armenians took over the Jewish quarter in Ankara, and then relieved when the Turks pushed the Armenians back across the border. They came to the United States because of all the fighting between the Armenians and the Turks. She and Victor Cohen were married in 1921. They had three children: Eli, Molly and Rachel. They also settled in the East New York section.

There is an interesting story concerning one of your probable ancestors. Many years ago Molly, Deborah, Victor and I visited the Mission in Carmel, California. On a plaque there we read that one of the founders was Father Crespi, who is also buried there. In a search on Google I read that Father Crespi had come from Majorca. When Molly's sister, Rachel, went to Majorca as a young woman in 1949 to visit where her ancestors had been, she discovered a large community of Conversos. Conversos were Jews who converted to Catholicism during the Inquisition to keep from being killed. To this date they are still known by that name. Father Crespi was a Converso.

Acknowledgments

I first wish to acknowledge my wife, Ruthie, for her insistence that I attend the memoirs class at the Los Altos Senior Center, for her encouragement and patience as I spent endless hours writing at the computer, and for her being my editor-in-chief.

I then wish to acknowledge Sylvia Halloran, my memoirs teacher, who created an atmosphere that drew me into her class and made me feel good about writing and reading my memoirs. Her corrections and recommendations are given in a manner that makes us feel good about ourselves. She encouraged all the members of the memoirs class to contribute to the critiques, which they do. Her memoirs class has become a close second family.

Thanks to Gail Ballinger for her many hours in helping me edit and organize this book. Thanks to Joan Garvin for her review, comments and corrections, which improved the final copy.

Thanks to members of my extended family and friends who encouraged me.

Last but not least, I would like to express my appreciation to my daughter Deborah and my son Victor. Their love, encouragement, help and caring, especially during difficult times, have made completion of my memoirs possible.

BIO MARCUS

Marcus, Bernard P.
Only in America : from
the South Bronx to

02/07/19

CPSIA information can be obtained
at www.ICGtesting.com
Printed in the USA
LVHW071412021218
598984LV00021B/890/P